T0188652

Governance &
Knowledge Management for
Public-Private Partnerships

Governance & Knowledge Management for Public-Private Partnerships

Herbert Robinson

Department of Property, Surveying and Construction
London South Bank University

Patricia Carillo

Department of Civil and Building Engineering
Loughborough University

Chimay J. Anumba

Department of Architectural Engineering
Penn State University

Manju Patel

NHS Grampian, Scotland

WILEY-BLACKWELL

A John Wiley & Sons, Ltd., Publication

Blackwell Publishing was acquired by John Wiley & Sons in February 2007. Blackwell's
publishing programme has been merged with Wiley's global Scientific, Technical, and Medical
business to form Wiley-Blackwell.

Registered office
John Wiley & Sons Ltd, The Atrium, Southern Gate, Chichester, West Sussex, PO19 8SQ,
United Kingdom

Editorial offices
9600 Garsington Road, Oxford, OX4 2DQ, United Kingdom
350 Main Street, Malden, MA 02148-5020, USA

For details of our global editorial offices, for customer services and for information about how
to apply for permission to reuse the copyright material in this book please see our website at
www.wiley.com/wiley-blackwell.

Library of Congress Cataloging-in-Publication Data

Governance & knowledge management for public-private partnerships / Herbert Robinson ...
[et al.].
    p. cm.
  Includes bibliographical references and index.
  ISBN 978-1-4051-8855-5 (hardback : alk. paper)   1. Communication in economic
development.   2. Public-private sector cooperation. 3. Knowledge management.
4. Construction industry–Management.   5. Public buildings–Finance.   I. Robinson, Herbert.
  HD76.G682 2010
  658.4′038–dc22   2009030794

A catalogue record for this book is available from the British Library.

Set in 10/12 pt Sabon by Aptara® Inc., New Delhi, India

1   2010

Contents

Foreword

Having delivered simultaneously one PFI programme and one PPP project in the early days when the initiatives were being used in the health sector, and before the rulebooks were written, I commend this book on 'Governance and Knowledge Management for Public-Private Partnerships'. The book focuses on two fundamental and interrelated themes; first, the need for effective project governance, and second, the necessity to capture and share lessons learned, which are both critical to the successful delivery of PFI/PPP projects. It is therefore an essential reading for new and existing practitioners in the wider field of capital investment and change management programmes and projects. The recommendation is heavily reinforced by my more recent experiences of organising and participating in Office of Government Commerce (OGC) Gateway Reviews in the health and central civil government arenas.

All sectors are sadly littered with programmes and projects that failed completely or at least failed to deliver the intended business benefits. If quantified, the reputational damage, waste of human effort, financial resource and the cost of the legacy would be seen as horrendous. The reasons were often due to a client that was inexperienced, inward looking and failed to prepare for the task they embarked on or unable to take the right decisions because of an inappropriate project management structure, inefficient processes and project teams not learning from previous mistakes.

Understanding the role of governance is critical for clients in developing effective project management structures and processes to facilitate good decision-making and to speed up the delivery of PFI/PPP projects. The lessons learned from the likes of National Audit Office (NAO) studies and OGC Gateway Reviews highlight continuing and repeated mistakes being made in project and programme delivery in all PPP/PFI sectors. These studies and many others have recognised the need for a knowledge management strategy supported by appropriate tools for capturing, sharing and applying knowledge to accelerate learning and build capacity for PPP/PFI projects.

The book not only gives a clear picture of the policy and strategic framework of PFI/PPP projects, the governance and knowledge management issues through different phases from planning, design development to operation and service delivery and the processes associated with each phase, but practical tools, methodologies and capabilities needed to deliver PFI/PPP projects in a range of sectors are also explained. It clearly demonstrates the key imperatives that are the hallmark of successful programmes and projects no matter what the method of funding or delivery.

I have no doubt that, in undertaking the programmes and projects that I am responsible for delivering and in undertaking OGC Gateway Reviews, to support clients delivering their own programmes and projects, I will draw on the material contained in this book.

Professor Rob Smith
Director of Gateway Reviews and Estates & Facilities
Department of Health

About the Authors

Dr Herbert Robinson (BSc Hons, MIP, PhD) is a Reader in Construction Economics and Project Management and the Director of the postgraduate programme in Quantity Surveying at London South Bank University. He teaches PFI/PPP projects as part of the corporate and project finance, procurement and management modules. He was involved in major research projects on knowledge management and knowledge transfer in PPP/PFI projects at Loughborough University. After graduating from Reading University with a degree in Quantity Surveying, he worked with international consultants Arup (United Kingdom) and in a World Bank funded public works infrastructure project in The Gambia before returning to academia. He holds a masters degree in Infrastructure Planning and a PhD in Infrastructure Investment and Resource Management. Dr Robinson is a member of the Editorial Advisory Board of the Journal of Financial Management in Property and Construction and has published widely on construction and PPP/PFI projects, knowledge and performance management. He is co-author of a book published on 'Infrastructure for the Built Environment: Global Procurement Strategies'.

Professor Patricia Carrillo (BSc, MSc, PhD, CEng, FICE, FCIOB) holds a personal chair in Strategic Management in Construction in the Department of Civil and Building Engineering at Loughborough University. She is also Programme Director for the department's MSc programmes in Construction Management and Construction Project Management. Her background is in civil engineering and she has worked with a range of clients, consultants and contractors. She was awarded the prestigious Royal Academy of Engineering Global Award and was also a visiting professor at the University of Calgary, Canada and University of Colorado, United States. Professor Carrillo's areas of expertise are in knowledge management, public-private partnerships, performance measurement and management, IT in construction and disaster resilience in the built environment. To date, she has published over 160 journal papers, conference papers and reports.

Professor Chimay J. Anumba (BSc Hons, PhD, DSc, Dr.h.c., PGCE, CEng, FICE, FIStructE, FCIOB, FASCE) is Head of the Department of Architectural Engineering at Penn State University, United States. He obtained his PhD in Civil Engineering (Computer Aided Engineering) from Leeds University (United Kingdom) in 1989 and has held a variety of positions in industry and academia. He has research interests in advanced engineering informatics, concurrent engineering, knowledge management, collaborative communications and project management. He has over 400 publications in these fields and has supervised more than 31 doctoral candidates and 15 post-doctoral researchers. Professor Anumba's work has received support worth over $15 million from various sources. He is co-editor of the *Journal of IT in Construction* (ITCon) and has held more than 10 Visiting Professorships. Professor Anumba has received numerous awards for his work including a higher doctorate degree (Doctor of Science, DSc) awarded by Loughborough University in July 2006 and an Honorary Doctorate (Dr.h.c.) by Delft University of Technology in The Netherlands.

Dr Manju Patel (BSc Hons, MBA, MSc, PhD and Fellow of IHEEM) is the Acute Sector Planning Lead, NHS Grampian, Scotland, United Kingdom. She holds a PhD in Neurosciences from University of London and later gained an MBA from City University Business School and an MSc with distinction in Construction Project Management (Healthcare). She has over 15 years experience working as a senior health service manager for various health organisations delivering strategic and operational changes and build solutions for major capital developments. Dr Patel was an internal health care adviser/consultant on one of the largest and most complex health care PFI projects in the United Kingdom. In a similar role, she produced a planning report for the incoming new NHS London in 2006; this document recorded the findings from a Pan London capacity and capital projects review. She has recently completed a 'Blue Print' health planning document for the acute services in NHS Grampian to support and inform all future build solutions on Foresterhill Site in Aberdeen, a shared Health Campus with the University of Aberdeen. She has presented numerous papers at national and international conferences and is a council member of the Institute of Healthcare Engineering and Estates Management (IHEEM).

Introduction

Public-private partnership (PPP) projects are significant given the demand for various type of partnership between the public and private sector, and the growing interest from other developed and developing countries to learn from the UK experience. Most of the previous books on PPP have focused on the procurement processes, examining specific issues such as risk management, legal aspects, finance and cost planning. However, governance and knowledge management in PPP projects have not been addressed. The aim of this book is therefore to fill that gap. First, by providing an understanding of the principles of governance and how it affects processes, people and actors. Second, to demonstrate how knowledge management can accelerate the learning and capacity building process to develop expertise and facilitate improvement in processes affecting planning and design development, construction and operational aspects of such projects.

This chapter starts with a brief context to provide an understanding of why PPP/PFI projects have become popular, and the key economic, technical and political arguments that have led to an increase in the use of PPP as a method of delivering traditional public services. The nature of such partnerships is defined and the specific features of PPP, as well as the different types of PPP projects, are explained. The evolution and development of PFI model in the United Kingdom are explained with specific reference to the impact of the Ryrie rules and the Bates reviews to improve the take-up and implementation of PPP/PFI projects. The role of governance and knowledge management to ensure continuous improvement in PFI/PPP projects is briefly explained followed by an organisation of the remaining chapters of the book.

1.1   The Context

There is a growing demand for investment to improve the quality of public services. Public sectors or governments worldwide are experiencing significant challenges as public resources are often insufficient to meet the increasing demand for new infrastructure projects to facilitate and sustain economic growth. As a result, there has been a growing and intense debate about the respective roles of the public and private sectors in the delivery of traditional public services. The United Kingdom and many other developed countries in Europe, United States, Canada, Australia, New Zealand and many developing and middle-income countries from Asia, Latin America and the Caribbean, Eastern Europe, Africa and the Middle East have now recognised the importance of the private sector in the delivery of traditional public services. There are a number of reasons for this. First, there are significant constraints in public sector investment affecting the quantity, quality and renewal of infrastructure stocks necessary to improve the delivery of public services and to enhance economic development. Second, there is evidence of poor performance in the execution and delivery of traditional public projects such as over-design, inadequate project and risk management resulting in time and cost overruns. The consequences are higher maintenance and operational costs associated with poor design and build quality. The traditional procurement approach of funding public sector projects also resulted in a huge backlog of maintenance leading to a deteriorating performance of infrastructure assets. The problems in the traditional procurement are exacerbated by a culture of separating the responsibilities for design, construction and operation. As a consequence, decisions on capital investment are often separated from operating expenditure critical for the effective maintenance and operations of public assets. Third, there is now an increased political will and awareness of the need to change by shifting the emphasis in public service delivery to outcome, results or performance-based approaches. However, it is widely recognised that an effective PPP policy and a strategic framework are required where the public sector is able to identify specific development needs, and engage the private sector to address them using their knowledge, innovation, technology, finance, technical and management skills.

The Private Finance Initiative (PFI) was introduced by the UK Conservative government in 1992 as a specific model of PPP. The Labour government consolidated the policy of encouraging private sector participation in the delivery of traditional public services in 1997 by developing a comprehensive PPP framework, with PFI as a cornerstone of the partnership. As Edwards and Shaoul (2003) noted, 'the delivery of state activities, which could not be privatised for financial or political reasons' are now transferred to the private sector under a range of partnerships. Such partnerships involved the public sector procuring the delivery of 'support' services and 'increasingly their core professional services' through long-term contractual arrangement in return for payment or fee from the public sector. Different types of PPP or variants of PFI were subsequently introduced such as Local Improvement Finance Trust (LIFT), Building Schools for the Future (BSF). However, they are underpinned by similar principles and the objectives of improving the delivery of traditional public services to ensure that the United Kingdom remains internationally competitive. For example, BSF was set up to 'transform the delivery of twenty-first century teaching and learning facilities in schools across England'. LIFT was set up to develop primary care and community-based health care facilities by creating a new market for investment through PPP.

Under the Labour government, there has been a significant momentum in the signing of PFI/PPP deals to modernise public infrastructure such as the underground, roads, hospitals, schools, housing and urban regeneration and other public buildings to improve services. However, the advent of new forms of PPP/PFI procurement has required a change creating fresh opportunities and challenges for both public and private sector organisations.

1.2   Key Drivers of PPP/PFI

Although the concept of PPP has existed for centuries in Europe, United States and other parts of the world, there are a number of reasons for the re-emergence of PPP/PFI projects in the United Kingdom. First, there is the central economic and efficiency argument focusing on value for money (Akintoye *et al*, 2003) and improving public services in health, education, transport and other core sectors in the United Kingdom. There are limits to the level of public expenditure available for infrastructure investment due to the constraints in public sector borrowing requirements (PSBR) as a result of Treasury and EU regulations (Fleming and Mayer, 1997). It was also a fundamental belief that the 'macroeconomic circumstances of the United Kingdom necessitate tight controls over public spending to restrain inflation' and to fulfil the Maastricht conditions to be part of the European Monetary Union (Grout, 1997; Broadbent and Laughlin, 1999). A key driver is therefore to facilitate infrastructure investment without imposing a heavy burden on public expenditure through the use of private funding (Forshaw, 1999). There is also the associated argument that the private sector is able to achieve greater efficiency and value for money in service delivery due to innovation to reduce whole life costs, risk management and the level of competition. The off-balance sheet argument has also been cited as a key factor. However, the Treasury argued that the objective of PFI is to provide high-quality services that represent value for money for the taxpayer and the key determinant of whether a project should go ahead is not the accounting treatment. PFI, therefore, eliminates significant capital expenditure requirements to design, build and own a capital asset. Instead, the public sector makes relatively small revenue payments (unitary charges) to pay for services delivered by the private sector throughout the concession period.

Second, there are the technical and environmental arguments relating to the traditional procurement process of delivering public assets. The fragmentation of design, construction and operation created a culture of focusing too much on minimising capital costs at the expense of whole life performance due to public sector practices separating capital expenditure from recurrent expenditure. It is increasingly argued that PFI can provide a valuable platform to improve the sustainability of buildings (Fell and John, 2003) due to its service-focused and whole life approach. The PFI approach is likely to result in the production of more efficient design solutions and functional buildings to minimise operational costs associated with maintenance and energy usage.

Third, there is the political motivation to urgently improve the level of public services such as reducing waiting lists in hospitals, tackling crime through urban regeneration and housing, and improving conditions in schools and the transport system. Voting outcomes are strongly influenced by the visibility of infrastructure projects and the level of public services. Under PFI, the initial expenditure requirement is considerably lower. As a result, infrastructure projects such as schools, hospitals, housing and roads unable to be funded using traditional procurement could go ahead. The public sector would therefore be able to undertake more projects with greater impact on public services which would otherwise have to wait longer to be implemented.

The economic and efficiency, technical and environmental as well as political arguments are central in the policy debate to stimulate the demand for PFI/PPP projects and to improve public services. A range of public sector or client organisations such as local authorities and NHS Trusts are therefore involved in PFI/PPP projects as it provides access to funding for large infrastructure projects and there is also a clear government policy to support it. It is therefore expected that the public sector will continue to be involved as long as they can engage the private sector and demonstrate value for money. The key drivers for the private sector are high returns on investment, profitability, steady stream of income in terms of unitary charge payments, long-term diversified workload and the opportunity to utilise their capabilities, competence and track record in PPP/PFI projects (Robinson *et al*, 2004). However, the continuous involvement of the private sector particularly design, construction organisations and other consulting firms depends upon getting good commercial returns, effective bid management to reduce cost and the level of political commitment.

1.3   Definitions and Key Features

A partnership is generally defined as a collaborative effort and relationship between parties to achieve mutually agreed objectives. However, a partnership involving the public and private sectors should be carefully structured to avoid potential problems because of the different value systems driving each side. Often, there is some tension between the private sector motive of profit maximisation and the public sector objective of delivering an acceptable level of service for public good in a manner that represents value for money. There are various definitions of the term 'public-private partnership (PPP)'. PPP is a generic term for any type of partnership involving the public and private sectors to provide services. It is generally a contractual arrangement where the private sector performs some part of a public sector service delivery responsibilities or functions by assuming the associated risks in return for payment. A recent research paper by the World Bank (2007) defines a PPP broadly as 'an agreement between a government and a private firm under which the private firm delivers an asset, a service, or both, in return for payments contingent to some extent on the long-term quality or other characteristics of outputs delivered'. According to HM Treasury (2000), 'public-private partnership' (PPP) is an arrangement that brings public and private sectors together in long-term partnership for mutual benefit. But regardless of the definitions, the objective is to utilise the strengths of the different parties to improve public service delivery and should always be underpinned by clear principles and contractual commitment reflecting a balance between profit and the need for regulation to ensure value for money in the use of public resources. For example, the private sector can reduce costs to increase profits through what Lonsdale and Watson (2007) refer to as quality shading to compromise service delivery to the public sector.

Under a PPP approach, public sector expertise are complemented by the strengths of the private sector such as technical knowledge, greater awareness of commercial and performance management principles, ability to mobilise additional investment, innovation, better risk management practices, and knowledge of operating good business models with high level of efficiency. PPP facilitate the exchange of skills between the public and private sector and improve the efficiency of resource allocation and the quality of public services. PPP programmes are therefore seen as an effective mechanism in delivering a long-term, sustainable approach to improve public services through investment, appropriate allocation of risks and rewards.

Partnerships are characterised by certain fundamental features. First, a partnership involves two or more actors or organisations, from the public and private sector which could also include the third sector, the so-called non-profit organisations. Sometimes the partnership is characterised by different types of private sector organisations complementing each other's role and interacting with different agencies in the public sector such as central and local governments resulting in complex relationships. Second, partnerships require some competitive element to select the best partner(s) and a degree of cooperation after selection (sometimes referred to as co-opetition). According to Lonsdale and Watson (2007), this is not a contradiction, as partners need to cooperate during the development of a PPP project and over operational matters such as monitoring and auditing service delivery. From the public sector perspective, 'competition is the best guarantor' for value for money (Broadbent and Laughlin, 2003). A third feature of partnerships is the existence of what is often referred to as an 'enduring and stable relationship' among the actors. This is achieved through cooperation, contractual obligations and commitment, once partners are selected through a competitive process. Fourthly, to fulfil their obligations, there are shared responsibilities defined by the contractual agreements for the resources and expertise required to achieve the project outcomes through specific delivery processes and activities. For example, planners, financiers, architects, engineers, surveyors, contractors and facilities management firms work together through various subcontracts to design, construct and manage a completed facility. Each party contributes resources to the partnership, in the form of money, land and skills to perform specific activities such as establishing the needs, appraising the options, planning and developing a business case for a project, evaluation of bids, design, construction, operation and maintenance. The roles in a partnership are formalised through various contract documents and the responsibility of each actor or partner is often reflected in interlocking subcontracts for design, construction, funding, cost and project management as well as facilities management. Another key feature is that the private sector is usually encouraged and given a high degree of freedom to provide innovative solutions that will represent value for the public sector based on the client's project or output specification. Finally, there is a risk–reward structure depending on the private sector inputs, requirements of the public sector and the service delivered by the private sector. The private sector party receives a fee or payment from the public sector usually based on predefined performance criteria and payment mechanism structured to reflect the risk allocation and incentives to avoid poor performance or quality shading. The payment may be entirely from service tariffs, or user charges or a public sector department's budget or a combination of both depending on the type of PPP.

1.4   Types of PPP/PFI Projects

PPP are implemented using different models. There are varying degrees of private sector composition and participation, resource allocation and risk–reward structure. The partnerships range from those dominated by the private sector to the other extreme where the public sector plays a dominant role. Different classification systems are used to categorise PPP/PFI projects based on investment, risk–reward structure, inputs or the range of specific activities involved. For example, HM Treasury (2007) identified three main types of PPP projects based on investment and reward structure such as financially free-standing projects, joint ventures and services sold to the public sector.

A *financially free-standing* PPP is where the private sector undertakes a project on the basis that costs will be fully recovered through user charges. The private sector recovers the capital expenditure involved in planning, designing, constructing an asset as well as the operating expenditure for operation and maintenance through, for example, a fee for using a toll bridge or road.

Joint ventures (JV) PPP projects are typically characterised by 'co-responsibility and co-ownership for the delivery of services' (Li and Akintoye, 2003). The projects are managed by the private sector, with the objective of delivering specific services to the public sector using their expertise, skills and finance. Usually, part of the project costs are recovered through some source of income other than payments by public sector such as tolls or other direct charges to users. The public sector contributes to achieve wider socio-economic objectives such as providing access and affordable transport, housing and other public facilities. Examples include infrastructure agreements for transport systems, housing and urban regeneration projects.

Service provision involves an arrangement where services are provided by the private sector to the public sector, typically by a Design, Build, Finance and Operate (DBFO) project. The public sector pays for the services provided by private sector through unitary charges or payments. Examples include privately financed social infrastructure such as health centres, libraries, schools and other forms of public or social infrastructure facilities.

PPP arrangements are also classified based on the range of activities required to deliver and manage the assets. It is important to distinguish between key activities relating infrastructure provision and production, the different roles played by the public and private sector. Traditionally, infrastructure provision relates to the planning, financing, monitoring and regulation of services. Production, on the other hand, focuses on the design, construction, maintenance and operation of the facilities. In the past, the public sector was responsible for both provision and production, but this was later followed by a gradual shift of production activities to the private sector due to privatisation in many countries. Such distinction between public and private sector activities is disappearing as there is now a growing trend to involve the private sector at every opportunity in the provision and production of public infrastructure. This point is illustrated by Edwards and Shaoul (2003), who noted that under PPP 'the government and its agencies are in effect becoming the procurer and regulator of services rather than the provider'.

Procurement of assets or services involves a range of interrelated activities from (1) planning, (2) financing, (3) design, (4) construction, (5) operation and maintenance to (6) monitoring and regulation of services (Howes and Robinson, 2005). Under PPP procurement, the public and private sectors participate based on the allocation of these activities to deliver an asset and to facilitate the delivery of core clinical or medical services (as in health care) or teaching services (as in education). There are therefore different PPP models as shown in Table 1.1 reflecting a combination of these key activities and requiring different types of payment regime such as usage, availability, operation and maintenance, and management fees (Aziz, 2007).

Under conventional procurement approaches, the public sector contracts out design and construction activities. The private sector carries out the design and construction tasks as separate activities as in the traditional 'architect-led' approach or as combined activities but with varying degrees of overlap as in 'design and build' and the management-based approaches such as construction management and management contracting. In Operate and Maintain (O & M) contract, the public sector outsource the operation and maintenance of the asset under a separate contract to the private sector after the facility is planned, designed and built as separate activities. The private sector is paid a fee for operating expenditure incurred in managing the asset but the public sector retains the responsibility for financing and ownership of the asset. Under other PPP models such as Design, Build, Operate and Maintain (DBOM) and Design, Build and Operate (DBO) procurement, significant activities are outsourced to the private sector. Some housing PFI projects are based on Rehabilitate, Operate and Transfer (ROT) model which involves rehabilitating existing asset owned by the public sector, managing the asset by operating and maintaining to a specified condition for a fee/payment during a period which is then transferred to public sector at the end. However, the most dominant and well-documented form of PPP in the United Kingdom is the DBFO model which underpins most PFI projects. Broadbent and Laughlin (2003) describes this as the 'exemplar PPP'. Through PFI, the responsibility for design, construction, operation and financing of infrastructure assets is transferred to the private sector usually for a period ranging from 20 to 30 years. The process involves creating an asset but the core objective is to deliver services to the public sector client in return for a performance-related payment reflecting the level of services provided. PFI is therefore a type of PPP which is fundamentally about the delivery of services rather than the procurement of assets (Birnie, 1999). A key feature of DBFO contracts is their long-term nature to allow for the economic amortisation of capital investment made by the private sector (Dowdeswell and Heasman,

Table 1.1   Types of PPP models.

Type of PPP model	Public sector responsibilities	Private sector responsibilities
Operate and Maintain (O & M)	Existing asset owned by public sector (already planned, designed, built and financed). Monitoring and regulation of FM services retained	Private sector manages the asset by operating and maintaining the asset to a specified condition for an operating and maintenance or management fee/payment
Rehabilitate, Operate and Transfer (ROT)	Existing asset owned by public sector transferred to the private sector. Planning/specifying the requirements for the assets/services	Private sector rehabilitates (involves modification of design/construction according to the specification/service requirements of the public sector and financing). Manages the asset by operating and maintaining the facility to a specified condition for a fee/payment during a period which is then transferred to public sector at the end
Design, Build, Operate and Maintain (DBOM)	Planning (specifying the requirements for the assets/services), financing capital cost of asset, monitoring and regulating asset/service performance	Designing the facility (subject to public sector requirements/specification), constructing, operating and maintaining assets (as well as financing the operating expenditure) for a fee/payment
Design, Build and Operate (DBO)	Planning (specifying the requirements for the assets/services), purchases asset for a pre-agreed price (financing). Monitoring and regulation of asset/FM services retained	Designing the facility (subject to public sector requirements/specification), constructing, operating and maintaining the asset for a fee
Design, Build, Finance and Operate (DBFO)	Planning (specifying the requirements for the assets/services), pay for availability and/or usage of assets (and services) through unitary charge. Monitoring and regulation of FM services retained	Designing the facility (subject to public sector requirements/output specification), financing, constructing, operating and maintaining the asset. Retains ownership and associated risks but assets transferred to public sector at the end. Receives a payment reflecting capital investment and operating expenditure

2004). There are also other variants which are widely used to reflect similar partnerships such as Build, Operate and Transfer (BOT), and Build, Own, Operate and Transfer (BOOT).

1.5   Evolution and Development of PPP/PFI

Privately financed infrastructure or PFI projects in the United Kingdom were initially subjected to a framework in the 1980s called Ryrie rules. There were two important elements to the Ryrie rules. First, privately financed projects should only be undertaken if value for money can be achieved compared to projects financed publicly. Second, this should also be accompanied by an equivalent reduction in public spending. The rules were set to prevent government departments from expanding and evading spending limits through private finance. However, the Ryrie rules were later criticised for being too restrictive and not providing incentives to pursue the private finance option. As a result, the rules were relaxed or modified in 1989 by eliminating the requirement for privately financed projects to be offset by an equivalent reduction in public spending. Subsequently, the Ryrie rules were fully retired in 1992 when the PFI was launched (Broadbent and Laughlin, 2003).

In 1992, the then Chancellor, Norman Lamont, announced the launch of PFI in the United Kingdom for the provision of public services, and to change government's attitude to privately financed infrastructure projects. However, few projects were signed partly due to limited knowledge and technical difficulties associated with this new form of PPP procurement. As a result of these problems and the slow start, a Private Finance Panel was created within Treasury as a knowledge centre to support PFI projects. There were a number of other changes to stimulate the use of PFI. For example, it was announced in November 1993 that public finance would not be available to NHS Trusts for capital investment without exploring the viability of the PFI route often referred to as universal testing. In November 1994, the NHS approach was adopted throughout, which meant that no public finance for capital projects would be approved unless PFI option does not provide value for money.

In 1997, Malcolm Bates, a former member of the Private Finance Panel, was asked to review the operation and delivery mechanism of PFI projects. The review concluded that the PFI should continue to be used by the public sector in partnership with the private sector to secure value for money. However, it was recommended that the public sector structures should be simplified and their roles and responsibilities clarified. A new Treasury Taskforce was subsequently established to develop PFI policy and to provide support for major projects. The Bates review also recommended the removal of the barriers affecting the progress of PFI projects. As a result, the Treasury Taskforce came together with government departments and the private sector to set the policy context and prepared technical notes which provided practical advice for implementing PFI procurement. This 'joined up' initiative resulted in, for example, the development of a standard template for PFI transactions. In 1998, the second Bates review recommended further changes to existing arrangements to improve the government's approach to PFI/PPP. There were also other changes suggested. For example, the HM Treasury paper, entitled 'PFI: Strengthening Long-Term Partnerships' (HM Treasury, 2006), identified ways in which the government could improve the PFI procurement process. This includes developing a secondment model within the public sector so that public servants with tacit knowledge or experience of complex procurements can be retained and deployed on projects across the public sector to facilitate the transfer of knowledge. This was further supported by the evidence of poor knowledge transfer for public capital projects as identified in the Green Public Private Partnerships Handbook (OGC, 2002).

1.6   Need for Governance and Knowledge Management

The changes to PFI procurement highlighted in the previous section reflect the importance of developing and applying new knowledge in continuously improving the delivery structure of PFI projects. The United Kingdom has experienced a steep learning curve and there have been a number of major reviews undertaken to improve the use of PPP/PFI for public service delivery. According to the World Bank (2007), the United Kingdom has driven much of the world thinking about PPP, and many other countries borrowed heavily from their experience in shaping their own PPP programmes. The increasing awareness of the successful application of PPP/PFI in addressing constraints in public funding in the United Kingdom has therefore resulted in a growing demand for this type of knowledge. Knowledge management is central to developing effective and sustainable PPP by accelerating learning and continuously improving PPP processes. Many other governments are now exploring private finance as an alternative means of funding to meet public service delivery needs. As a relatively new form of PPP, there are important lessons learned from the UK experience in PFI/PPP projects that can be transferred to other countries, particularly where there are budgetary constraints and the need to improve the level of public services is greatest.

PFI/PPP projects are required to represent value for money (VFM) when measured against an equivalent project delivered through traditional public funding. However, the VFM argument to establish the need for a PPP project places high expectation on the ability and knowledge of *people, actors in government departments* in the partnership and the efficiency of the *processes* used to deliver projects. It is therefore essential to understand the governance mechanisms, firstly, to control the actions of people and actors in government departments to observe due processes, and secondly, to accelerate learning to develop expertise and improve processes in PFI/PPP delivery. PFI/PPP projects, therefore, require an effective governance framework of processes and controls for people's actions and government actors to safeguard against poor decision-making, error and fraud, illegal transactions resulting in inappropriate delivery, poor VFM or project failure.

Governance is the 'the act, manner or function of regulating the proceedings of a corporation' or simply to steer, exercise restraint, control the speed and actions, policies or affairs of a nation, an organisation or project. Good governance in terms of people and actors (soft) and processes (hard) is therefore critical for the successful delivery of PFI/PPP projects. As a relatively new form of procurement, there are shortages of PFI/PPP experts. It is therefore important that lessons learned about processes and tacit knowledge of people involved are codified or transferred effectively to other individuals or organisations interested in PFI/PPP projects. This is absolutely critical for countries where there are public sector budgetary constraints and the need to improve the level of public services through PPP.

Understanding the role of governance and how to transfer lessons learnt through knowledge management and capacity building is fundamental to facilitating a sustainable improvement in the delivery, efficiency and effectiveness of PFI/PPP projects. Knowledge gained by the authors from recent research on knowledge management, PFI/PPP projects and governance in PFI projects has identified the need for a better understanding of how governance and knowledge management can facilitate the delivery of projects. The findings from the research has underlined the significance of both governance and knowledge management; hence, the book is aimed at bringing together two of the most important aspects of good governance and the transfer of lessons learnt to continuously improve PFI/PPP delivery. The book focuses on how to improve *processes* and the decision-making ability and expertise of *people* and actors in PFI/PPP transactions using a governance and knowledge management approach to ensure a successful project outcome.

1.7   Organisation of the Chapters

Following this introduction, the book is divided into four parts. Part 1 (Chapters 2 and 3) starts by examining the policy and strategic context to provide an understanding of key policy and strategic variables, nature of PFI/PPP projects, the key principles underpinning PPP, structure and the delivery mechanisms of PFI/PPP projects. This part provides the context for the subsequent chapters in Part 2 on the principles of governance and its application, and Part 3 on knowledge management theories, principles and practices. Part 4 focuses on the need to improve governance and knowledge management through capacity building and a framework for knowledge transfer and learning.

Part 1: Policy, Strategy and Implementation

Chapter 2 focuses on the policy and strategic considerations for PFI/PPP projects. The key policy elements of PFI/PPP such as policy theory and objectives, the institutions and their roles, expertise and resources, processes, information and knowledge management systems, monitoring and evaluation as well as the policy environment are examined. The governing principles of PFI/PPP projects such as VFM, risk transfer, whole life commitment, focus on core services and payments based on performance underpinning PFI theory are outlined. The management structure and strategy is discussed in terms of the team composition, contract and interface management between key stakeholders, the need for stakeholder analysis to identify potential impact on others affected outside the core group, the key benefits, expectations and risk faced by different stakeholders to develop effective and successful PFI/PPP projects. The funding strategy of PPP/PFI projects, the importance of bankability, project structuring and credit enhancement to make PFI/PPP projects attractive, strengthen risk management and the viability are also explained. Whole life assessment and the need to integrate sustainability strategy are discussed. There is also some reflection on PFI/PPP projects from a European and international perspective identifying key regions, countries and the level of investment, in particular PFI/PPP sectors and market.

Chapter 3 discusses the delivery phases of PPP/PFI projects. The key stages and issues associated with the procurement of PPP/PFI projects from planning and design development to construction, operation and service delivery are examined and discussed. The planning and design development phase is identified as crucial for the success of PPP/PFI projects. Key issues relating to needs assessment, developing a business case, and advertisement to generate interest and create competition necessary to achieve VFM are discussed with respect to the competitive dialogue and negotiated procedures. The importance of the output specification which provides the basis for design by defining the standards for accommodation/facilities and the scope of hard and soft facilities management services required by the public sector client and its role in justifying a PFI/PPP solution and determining affordability is explained. The specific issues relating to invitation and pre-qualification of potential bidders, design development, evaluation of bids, selection of the preferred bidder, financial close and developing the full business case for the PPP/PFI project are also examined. The construction phase focusing on the assembling and production process and key issues relating to phasing of completed projects and decanting are examined. The operation and service delivery phase involves the management of the completed facilities for service delivery. From the public sector client perspective, the operation and maintenance phase is the most crucial, so the role of performance monitoring and payment mechanisms to ensure VFM is achieved is discussed.

Part 2: Concept, Principles and Application of Governance

Chapter 4 examines the principles of governance and how they relate to key issues at various phases and stages in PPP/PFI projects discussed in Chapter 3. It starts with a review of the concept, principles and dimensions of governance. The objectives of governance to control processes, decision-making, and behaviour of people and actors in public sector to ensure project outcomes are not hindered or compromised are explained. Key components of governance such as project approval, procurement processes, control mechanisms such as standards or procedures, organisational structures, accountability and post-project evaluation are related to the key phases and stages of PPP/PFI delivery. The importance and the role of Gateway Review at key phases and stages in the delivery process are highlighted and discussed. Following the adoption of PPP/PFI, various governance tools have been established such as a business case to assess the compliance of the completed project to its original objectives, a project team with roles and responsibilities clearly assigned, a defined method of communication to each stakeholder, an agreed specification, a plan that spans all stages from initiation through to completion and managing risks identified during the project are discussed. The relationships between internal and external stakeholders involved in the project, the flow of project information to all stakeholders and the approval mechanisms at appropriate stages of PPP/PFI projects to monitor compliance are analysed.

Chapter 5 uses case studies from the health sector which is one of the most significant, complex and mature sectors in terms of the level of investment and development of PFI/PPP in the United Kingdom to reinforce concept and principles of governance discussed in Chapter 4 and to assess its impact on project delivery. Four recent PFI projects are selected with varying degree of complexity, organisational, development and implementation challenges. The first two (Case Studies 5.1 and 5.2) are simple early wave PFI schemes, built on demolished or adjacent brownfield land. The other two (Case Studies 5.3 and 5.4) are highly publicised and complex PFI schemes involving mergers or co-location of more than one National Health Service (NHS) Trust's onto single or multiple sites. Case Study 5.3 examines the early planning phase of a complex multi-organisation PFI scheme, involving co-location of two NHS Trusts and a single research institute onto a single site. Case Study 5.4 examines the early planning phase right through to the completion of the full business case (FBC) for a complex, single NHS Trust but multi-site PFI scheme. Findings relating to project governance focusing on key issues such as reporting structure and levels of responsibilities, project controls, risk management, and critical success factors at various stages of project delivery are compared and discussed. The key similarities and differences in each case study organisation's approach to project governance and the relationship with project delivery in terms of success and failures are summarised, analysed and discussed. The lessons learnt and the need for knowledge transfer is identified as crucial to improve governance and the performance of future PPP/PFI projects.

Part 3: Theory, Principles and Application of Knowledge Management

Chapter 6 focuses on the theory and principles of knowledge management; the different types of knowledge and dynamics of knowledge are explained using Nonaka and Takeuchi's model of knowledge creation. The key building blocks and elements required to develop a knowledge management strategy such as knowledge management goals, dimensions of knowledge, leadership, resources and the strategic options available for effective knowledge management implementation for PPP/PFI projects are discussed. The application of practical tools developed in collaboration with leading design and construction firms for implementing KM strategy and benchmarking KM implementation efforts in project organisations such as CLEVER, IMPaKT and STEPS are described to show how knowledge can be managed effectively to improve the performance of PPP/PFI projects.

Chapter 7 uses case studies from public sector client organisations and leading private sector organisations involved in PFI/PPP projects to capture the perspectives of various stakeholders in PFI/PPP. Four case studies are selected reflecting the experience of the Public Sector Client (Case Study 7.1), Special Purpose Vehicle (Case Study 7.2) Consultant and Adviser (Case Study 7.3), Design and Build Contractor and Facilities Management Provider (Case Study 7.4). The role and activities of the case study organisations, types of knowledge required and key issues at critical stages of delivery such as the outline business case, preferred bidder, facilities management (FM) and operational stages are examined. The key problem areas and scope for learning to acquire the knowledge required to continuously improve processes and the decision-making ability of actors in the public and private sector organisations are discussed. The knowledge transfer issues in PFI/PPP projects, the implications for various stakeholders in terms of the improvement capability and organisational readiness of organisations to adopt a knowledge management strategy are also examined. The need for the development of a practical framework to facilitate the development of knowledge transfer capabilities in PPP/PFI projects is identified as crucial to accelerate learning and capacity building.

Part 4: Knowledge Transfer and Capacity Building

Chapter 8 focuses on the key issues and some of the challenges in building capacity to accelerate the delivery and improve the performance of future PPP/PFI projects. Approaches for developing explicit and tacit knowledge are examined through the development of best practice, guidance documents and knowledge centres such as Partnerships UK, 4Ps, PPP dedicated units, Office of Government Commerce and Treasury. The role of training and capacity building institutes, research and development, technical assistance and advisers to improve governance and to facilitate knowledge transfer in the implementation of PPP/PFI projects are also examined.

Chapter 9 discusses the application of a practical tool/framework evaluated by industry partners to demonstrate how to implement a strategy to accelerate learning and capacity building process for organisations involved in PFI/PPP projects. The key development stages or steps for using the knowledge transfer framework are identified and discussed. The three-stage framework involves (1) improving participation and exploring opportunities in PFI/PPP, (2) building a knowledge map and transfer capability and (3) implementing a knowledge transfer action plan to facilitate improvement in PFI/PPP projects. The evaluation of the knowledge transfer framework and the benefits to PPP/PFI organisations are also discussed.

Chapter 10 is the concluding chapter providing some reflection on current issues and challenges relating to governance and knowledge management affecting key phases from planning and design development to construction and operation of PPP/PFI projects. The need for an output specification to clearly define the requirements of the public sector and a performance monitoring mechanism to ensure continuous improvement are highlighted as crucial and the implications for the sustainability of PPP/PFI projects are outlined.

References

Akintoye, A., Hardcastle, C., Beck, M., Chinyio, E., and Asenova, D. (2003) Achieving best value in private finance initiative project procurement. *Construction Management and Economics* 21, 461–470.

Aziz, A.M.A. (2007) A survey of the payment mechanisms for transportation DBFO projects in British Columbia. *Construction Management and Economics* 25, 529–543.

Birnie, J. (1999) Private Finance Initiative (PFI): UK construction industry response. *Journal of Construction Procurement* 5(1), 5–14.

Broadbent, J., and Laughlin, R. (1999) The private finance initiative: clarification of a future research agenda. *Financial Accountability and Management* 15(2), 95–114.

Broadbent, J., and Laughlin, R. (2003) Control and legitimation in government accountability processes: The Private Finance Initiative in the UK. *Critical Perspectives on Accounting* 14, 23–48.

Dowdeswell, B., and Heasman, M. (2004) *Public Private Partnerships in Health: A Comparative Study*. Report prepared for the Netherlands Board for Hospital Facilities. The EU Health Property Network with The Centre for Clinical Management Development, University of Durham.

Edwards, P., and Shaoul, J. (2003) Partnerships: for better, for worse? *Accounting, Auditing and Accountability Journal* 16(3), 397–421.

Fell, D., and John, R. (2003) *Sustainability Incentives: Contracts and Payment Mechanisms, CIEF (Construction Industry Environmental Forum?) Meeting Notes 25th November*. Commonwealth Institute, London.

Fleming, J., and Mayer, C. (1997) The assessment: public sector investment. *Oxford Review of Economic Policy* 13(4), 1–11.

Forshaw, A. (1999) The UK revolution in public procurement and the value of project finance. *Journal of Project Finance* 5(1), 49–54.

Grout, P.A. (1997) The economics of the private finance initiative. *Oxford Review of Economic Policy* 13(4), 53–66.

HM Treasury (2000) *Public Private Partnerships: The Governments Approach*. The Stationery Office, London.

HM Treasury (2006) *PFI: Strengthening Long-Term Partnerships*. HM Treasury, London.

HM Treasury (2007) *Accounting for PPP Arrangements Including PFI Contracts*. Financial Reporting Advisory Board, Discussion Paper, December. The Treasury, London.

Howes, R., and Robinson, H. (2005) *Infrastructure for the Built Environment: Global Procurement Strategies*. Elsevier Butterworth-Heinemann, Oxford.

Li, B., and Akintoye, A. (2003) An overview of public-private partnership. In: Akintoye, A., Beck, M., Hardcastle, C. (eds). *Public Private Partnerships: Managing Risks and Opportunities*. Wiley-Blackwell, Oxford, pp. 3–30.

Lonsdale, C., and Watson, G. (2007) Managing contracts under the UK's Private Finance Initiative: evidence from the National Health Service. *Policy and Politics* 35(4), 683–700.

Office of Government Commerce (2002) *Green Public Private Partnerships*. OGC, London.

Robinson, H.S., Carrillo, P.M., Anumba, C.J., and Bouchlaghem, N.M. (2004) *Investigating Current Practices, Participation and Opportunities in Private Finance Initiative*. Loughborough University, Leicestershire.

Sustainable Development Department in East Asia and Pacific, The World Bank (2007) *Public Private Partnership Units: Lessons for Their Design and Use in Infrastructure*. The World Bank and Public-Private Infrastructure Advisory Facility (PPIAF).

2

Policy and Strategic Framework

2.1   Introduction

Public-private partnerships encourage long-term relationships between public sector and private sector organisations to facilitate the provision of schools, hospitals, transport and other essential public services. Since the introduction of the PFI/PPP model in the United Kingdom, private participation and investment in public infrastructure projects have increased significantly. However, there was a delay in the initial take-up of PFI schemes due to a number of problems relating to clarity of policy objectives, processes and delivery mechanisms, institutional, legal and capacity constraints, and limited knowledge. For example, in the health sector, there were concerns relating to competition issues and the risk of local substitution after major PFI investment because of the internal market created. In housing, there was a need for changes in the legislation governing the way local housing authorities arranged for other agencies to carry out their housing management functions by entering into management agreements.

This chapter focuses on the need for a policy and strategic framework for PFI/PPP projects. Following this introduction, the key elements of the policy framework such as policy theory and objectives, institutions and their roles, expertise and resources required, processes, information and knowledge management systems, monitoring and evaluation mechanism as well as the policy environment are examined. The governing principles underpinning PFI theory such as value for money and risk transfer, whole life approach, facilitating the delivery of core public services and payments based on performance are then discussed. The need for an effective strategy relating to the management structure and stakeholder engagement to develop effective and successful PFI/PPP projects are also examined. The funding strategy in terms of the capital structure, the importance of project structuring, bankability, and credit enhancement to make PFI/PPP projects attractive, strengthen risk management and the viability as well as the importance of a sustainability strategy are also discussed. There is some reflection on the European and international PPP market as it will have a major influence on companies strategy.

2.2   The Policy Framework

Decisions on the nature of public services required and how they should be provided are fundamental in public policy. The effectiveness of the policy framework influences the outcome of a policy. However, this depends on key ingredients such as the *policy theory and objectives* defining the types of services and outcomes expected, the *institutions* involved and their roles, *expertise and resources* available in the public and private sector, *processes*, and *monitoring and evaluation mechanisms* for control and to assess actual outcomes. *Information and knowledge management systems* are required to promote opportunities, share experience and feedback or feed 'forward' lessons learnt in PPP/PFI projects. The *policy environment* to support the development and implementation of PPP/PFI is also crucial. Figure 2.1 shows the relationship between the key elements in the policy framework.

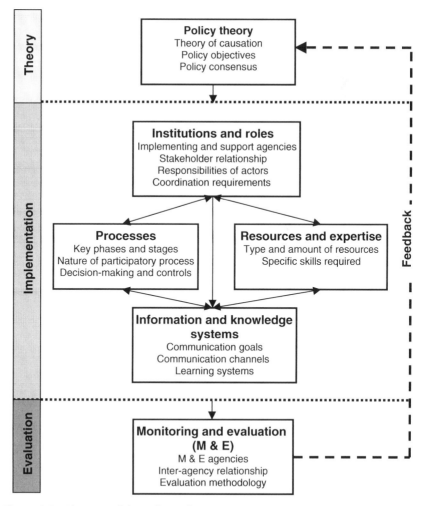

Figure 2.1   Elements of the policy making process.

2.2.1   Policy environment

Understanding the policy environment is fundamental in developing and implementing policies. There are various stakeholders involved in PPP/PFI projects with a range of views and interests. This includes public sector actors such as politicians, civil servants in government departments, public sector advisory and regulatory agencies, private sector participants such as investors and lenders, design and construction firms, users and special interest groups such as trade unions. It is important to recognise how the interests of different stakeholders, which are sometimes conflicting, may have an impact on the success or failure of a policy. In developing PPP/PFI policies, there is a need to carefully consider the views of the private sector stakeholders and the marketplace to assess support for such projects.

The nature of the relationship between the public and private sector organisations depends on how PFI/PPP projects are delivered. The market environment will determine the nature of competition and level of interest, the scope of PPP/PFI scheme, and the responsibility of the public and private sectors. Issues relating to who is likely to participate, how the scheme will be structured, how services provided will be monitored, the form and duration of the contract and the risk–reward structure should be carefully considered. PPP/PFI projects should therefore be designed to be attractive and commercially viable in order to secure the participation of the private sector, to reflect the market environment and specific sector characteristics, otherwise projects are likely to fail.

Political support is crucial to build momentum and maintain confidence in PPP/PFI schemes. More significantly, it helps to ensure that the concerns of the private sector in the marketplace and other stakeholders are adequately addressed. Clarity of policy objectives is strongly influenced by the political environment, and is likely to have a significant effect on the level of public support, private sector participation, and therefore the levels of private investment. In the United Kingdom, both the main political parties, Labour and Conservative, are committed to policies and the principles of private finance to support public services. There is likely to be future re-branding such as LIFT, Building Schools for the Future and other models on public-private partnerships to reflect lessons learnt from earlier schemes. There will be some adjustments and policy refinement or modifications needed as a result of the development of new knowledge for continuous improvement.

2.2.2   Policy theory and objectives

All policies are based on a concept of moving from one particular situation to a desired state, and every policy implies a theory or causal relationship. The challenge for policy makers is to convert theory into policy objectives and beneficial outcomes. There is considerable debate on the theory, rationale and justification of PPP/PFI policies ranging from transfer of risk to the private sector, value for money, increasing private investment, off-balance sheet financing, innovation to whole life approach in the delivery of public services. However, the overall objective of PFI/PPP is to improve the level of public services. The PFI model is underpinned by a theory focusing on the delivery of services rather than the ownership of assets (Grout, 1997). The contestability of public service delivery is therefore at the heart of the theory and the key question is whether the level or quality of services can be provided or delivered cheaper by the private sector. The principal–agent theory provides a good theoretical framework for understanding service contracts (Martin, 2004). The principal's objective (public sector client) is to maximise utility by demanding level of services stipulated in the contract and reducing transaction costs associated with monitoring to achieve practical value for money. The agent's objective (PFI contractor), on the other hand, is to maximise utility (e.g. profit and reputation) by increasing performance-related payments and avoiding deductions for service failures. Service contracting makes economic sense if the private sector contractor or agent's *production* and *transaction* costs are less than in-house costs of the public sector client or principal. Production cost relates to the whole life cost of producing the asset and delivering services, and transaction costs are incurred by the private sector contractor in bidding and negotiating for the service contract. The public sector client (principal) also incurs transaction cost in obtaining information for setting performance standards, monitoring private contractors and negotiating during contract service delivery.

An effective policy formulation requires two things. First, understanding the nature of the problems, and identifying the relevant theory or theories relating to how the problems can be addressed. The traditional procurement approach is fragmented and has resulted in poorly performing or dysfunctional buildings delivering poor services. PFI theory addresses these problems by creating a shift in emphasis from 'building contracting and lump sum payment' to 'service contracting and performance-based payment'. Second, determining what elements are required to implement and achieve the desired policy outcome in terms of institutions and their roles, processes, expertise and resources, information and knowledge management systems, and a monitoring and evaluation mechanism to assess the impact of policy objectives. The policy objectives agreed should be the result of consensus between different stakeholders in the public sector and the private sector organisations.

2.2.3   Monitoring and evaluation

An appropriate monitoring and evaluation mechanism is a critical aspect of policy development to control the behaviour of actors, and to avoid poor decision-making so that undesired consequences are eliminated or minimised during the different phases and stages of the PPP/PFI project. Monitoring and evaluation mechanisms are needed to reduce potential abuse from key actors including the private contractor (agent) to ensure that the intended policy outcomes are achieved and the public sector client (principal) gets value for money in the delivery of services.

Under PFI contracts, the public sector client (principals) delegates the function of service delivery to PFI/PPP contractors (agents) based on output specification defining the contract or scope and objectives for service delivery. Powerful incentives and penalties are used on the one hand to transfer risks, with performance monitoring involving a degree of cooperation on the other hand to measure compliance with output specifications to ensure that value for money is achieved. The objectives of both parties is to maximise utility but the relationship between public and private sector actors will have a significant impact due to the long-term nature of the services delivered. The PPP/PFI contractors (agents) have more knowledge and information about their daily service delivery activities than the public sector (principals). Performance monitoring in PFI projects is therefore critical as the 'agent may abuse its information superiority to maximise its own utility' (Wang *et al.*, 2007). The principal's utility may be jeopardised if service delivery or the behaviour of the agent is difficult to monitor. The ability of the principal (public client) to obtain day-to-day information on service delivery depends on the willingness of the contractors (agents) to provide information and the ability of the principals to independently discover information through different performance monitoring regimes.

Standard agency theory shows that an optimal incentive contract involves a performance-based and payment system linked to the agent's performance score which is assumed to correlate strongly with the agent's effort level. The knowledge developed through monitoring and evaluation of existing PFI/PPP projects should therefore be communicated properly to facilitate a continuous improvement in PFI/PPP policies and delivery mechanisms.

2.2.4   Institutions and roles

The institutional framework defines the implementing agencies and their roles, and in PFI/PPP projects, both the public and private sector organisations are involved. The actors on the public sector side include local authorities, National Health Service Trusts, various government departments, support and advisory agencies such as 4Ps and Partnerships UK. Table 2.1 provide examples of public sector departments and support agencies with clearly defined roles. It is important to have a central PPP policy unit to guide and direct the implementation of PFI/PPP projects. Coordination requirements between the different government agencies are also vital to speed up the PPP/PFI process. There is therefore a need for strong coordination between the Treasury Department, Office of Government Commerce (OGC), National Audit Office (NAO) beneficiary government departments particularly the Departments of health, education, transport and local authorities with major projects and support agencies notably 4Ps and Partnerships UK. Partnership UK and 4Ps have merged to form a new joint venture called 'Local Partnerships' with effect from August 2009 to work with local public bodies in supporting the improvement of public services and infrastructure.

The private sector actors include consulting architectural, engineering, surveying firms, building and civil engineering contractors, banks, financial, legal and other specialist technical advisers. The private sector actors form a special purpose company (SPC) or vehicle (SPV) to deliver a PFI/PPP project and usually employ external financial and legal advisers where specialist

Table 2.1 Public sector organisations and their roles in England.

Institution	Role
Treasury Department	The government created an *operational taskforce*, acting on behalf of HM Treasury, based in Partnerships UK. The taskforce set up a help desk to assist public sector partners with operational PFI issues. Provides update on all PFI/PPP projects signed and responsible for overall PFI/PPP finances, expenditure control and management
Office of Government Commerce (OGC)	An independent office of the HM Treasury providing procurement advice
Project Review Group (PRG)	The PRG oversees the approval process for local authority PFI projects that receive government support. It is the gatekeeper for the delivery of PFI credit funding to the local authority PFI programme
Partnerships UK (PUK)	Partnerships UK (PUK) is a PPP agency which has a unique public sector mission: to support and accelerate the delivery of infrastructure renewal, high-quality public services and the efficient use of public assets through better and stronger partnerships between the public and private sectors
National Audit Office (NAO)	1. Audit the accounts of all government departments and agencies as well as a wide range of other public bodies 2. Report to Parliament on the economy, efficiency and effectiveness with which these bodies have used public money
Strategic health authorities	Capital Investment Unit provides support to NHS Trusts in developing PFI/PPP schemes
Local authorities	Provide an 'area wide vision', strategic business case for school provision or Strategy for Change (SfC) to outline estate strategy and objectives of capital investment working closely with PfS team
Partnership for Schools (PfS)	Set up as the delivery agency for BSF working with local authority and private sector. Jointly managed by the then DfES Department for Education and Skills and PUK. Approval for BFS
4Ps	Assists local authorities in the development, procurement and implementation of PFI/ PPP projects and other contractual partnering arrangements

knowledge is required. The SPC/SPV can be a specially formed subsidiary of an existing construction company, a joint venture, a consortium or a specialist PFI company.

The nature of the relationships between public and private or inter-agency relations are crucial in PPP/PFI projects. A key challenge for the public sector is to define the service requirements or outputs. The private sector, on the other hand, will need to interpret the public sector requirements, develop a building solution and understand their service obligations or responsibilities with respect to the performance of assets and service delivery. The PFI approach has clearly shifted and increased the risks on the PFI contractor as 'liability is inevitably extended under performance-based contracts' (Gruneberg *et al.*, 2007). Unlike traditional procurement, where the liability of the contractor is normally restricted to a shorter defects liability period, usually 12 months, the PFI contractor is liable not only for asset performance but also for a wide range of hard and soft facilities management (FM) services during the contract period of 20–35 years. The responsibilities are profound for private sector bidders in terms of managing the relationship between the different actors involved in PFI/PPP projects to achieve the contractual level of service performance.

Poor coordination between the private and public sector can sometimes lead to excessive delay in the implementation of PFI/PPP projects. The number of different institutions and actors involved could have an impact on the level of coordination required if relationships are not managed properly. Whilst it is important to have appropriate support, checks and balances to prevent abuse and wrong decision-making, too many institutions and actors with diverse perspectives and interests serve to increase the problems of coordination which could muddle policy objectives.

There are also other supporting roles within various government departments. For example, the Department of Health provides specialist private financing guidance via its Private Finance Unit. To date, over 20 government departments in the United Kingdom have benefited from PFI/PPP schemes with significant investment particularly in the health, education and transport sectors, prisons, fire stations, waste management housing and urban renewal.

2.2.5   Expertise and resources

Projects procured under the PPP/PFI approach require significant private investment and expertise. Skills are required for planning, design, construction, operation and maintenance of completed facilities and for monitoring the services provided. In the public sector, specialist expertise is required for project initiation, needs assessment, options appraisal and developing a business case for PFI projects. If specialist expertise is not available, technical, financial and legal advisers will be required. Technical advisers include planners, architects, engineers and quantity surveyors, asset and facilities managers and other specialists dealing with all aspects of planning, design, construction and operation. For example, this may include health, transport, education planners and epidemiologists. Equally, it is important to have the range of expertise or skills in private sector firms or the consortium. Highly specialised knowledge required to undertake different tasks in PFI/PPP programmes, for the public and private sector, should be carefully assessed in terms of skills set required as part of the policy development process. Lack of skills could seriously derail the implementation of PFI/PPP programmes. In the United Kingdom, for example, the lack of contractors specialising in complex hospital PFI/PPP projects threatened to undermine the level of competition required to achieve value for money.

The type of skills and investment required is crucial particularly at the early stages of implementation. In transition and developing economies where resource markets are often underdeveloped and unpredictable, there could be significant increases in transaction and infrastructure development costs due to shortages of various technical expertise. The funding implications of PFI/PPP schemes should also be assessed to determine public sector obligations in terms of future expenditure required such as regular payments/unitary charges for signed PPP/PFI projects. The level of private sector interest, their capacity to attract private finance both debt and equity as well as the willingness to participate in long-term partnership with the public sector will have a significant influence on the level of investment available. It is therefore essential that sufficient investment (from the public and private sectors) is available to support the implementation of PPP/PFI projects. Treasury data and information show that as of end February 2009, there were about 630 signed projects and 540 operational PFI deals with a total capital value of over £63 billion (HM Treasury, 2009). Over 34 hospitals, 239 new or refurbished schools and other public infrastructure facilities such as transport, prisons, housing and accommodation schemes are already up and running.

2.2.6   Processes

Developing and implementing successful PFI/PPP projects require well-defined and robust processes to facilitate decision-making, to prevent abuse and safeguard public resources from the planning and design development phase, construction to operation and service delivery phase. Processes are required for developing business cases, selecting public sector advisers, reviewing and approving schemes, advertising and market testing to ensure private sector interest, assessing the commercial viability of projects, inviting and submitting bids, evaluation of bids, negotiation, participating in dialogue and the selection of a preferred bidder. For example, preparing a business case at the early planning and development stages requires a clear understanding of specific steps and processes, all of which can be found from a number of guidance documents developed by client departments such as the Department of Health and the Treasury. The Capital Investment Manual or CIM forms the main reference document for both the outline and full business case stages of the capital planning process, and since the first issue in 1994 (DoH, 1994), the manual has recently been supplemented, to include privately financed schemes. The Treasury has, for many years, provided guidance (i.e. Treasury Green Book, 1997, with revision in 2004) to public sector bodies on how proposals should be appraised, before significant funds are committed. The revised edition (April 2003) of the Green Book is designed to encourage a more thorough, long-term, analytically and robust approach to appraisal and evaluation. To this end, the two main changes introduced are lowering the previous discount rate of 6% to the new 3.5% and the requirement to make an adjustment in appraisals for optimism bias (i.e. pricing for risks uncertainties). Both changes are viewed as making business cases more robust and realistic in terms of value for money. The third guidance, 'Principles of Generic Economic Model for Outline Business Case Option appraisal' (DoH, 2004), serves to help with the financial appraisal of the shortlisted options, and as such, at each consideration of a principle, the guide appropriately links to the Excel OBC Model. This guidance should be valued for its simplicity in that it is written to serve the much wider audience than specific sectors, financial accountants/advisers. The document itself provides clear guidance on the key economic concepts and principles;

Table 2.2   Examples of key documents to facilitate process.

Institution	Examples of key documents
Treasury Department	Operational Taskforce Note 1: Benchmarking and Market Testing Guidance. Guidance designed to support public sector PFI contract managers in achieving value for money through benchmarking and market testing of soft services
	Operational Taskforce Note 2: Project Transition Guidance. This guidance is designed to support project and contract managers in the transition from procurement to operation
	Operational Taskforce Note 3: Variations Protocol for Operational Projects (entered into prior to Standardisation of PFI Contracts version 4). This protocol is to help public sector authorities with PFI contracts to put in place a voluntary protocol for managing variations during the operational phase of PFI projects
OGC	Information on Government Procurement Service; Gateway Review Booklets; Best Practice Guidance and PRINCE 2
PRG	Process and Code of Practice Evaluation Framework – Guidance to Reviewers
Partnerships UK	Guidance Notes on the following: HMT PFI Guidance: Standardisation of PFI Contracts Version 4 (SoPC4) HMT PFI Guidance: Standardisation of PFI Contracts Version 3 (SoPC3) HMT PFI Guidance: Change Protocol Principles HMT PFI Guidance: SoPC4 Drafting Pack for Updating Contracts HMT PFI Guidance: Value for Money Assessment Guidance
NAO	The NAO produces technical information, guidance and good practice material that can be used by others, particularly those in public sector organisations
Strategic health authorities	Capital Investment Manual Standard Output Specifications

it describes not only how these are used in economic appraisals but also how the appraisals are interpreted.

For key stages of PFI procurement, specific guidance and technical notes reflecting different sub-processes should therefore be available to help public sector organisations and their advisers to explore the potential for PFI, make decisions within their organisations on how to implement PFI/PPP projects. A number of PFI/PPP documents/publications are regularly produced, updated and continuously reviewed by the government and support agencies such as OGC, 4Ps (Public-Private Partnership Programme) and Partnerships UK to demonstrate or explain how to put PFI theory into practice and to improve the implementation process. Examples of some documents to facilitate and improve processes are shown in Table 2.2.

2.2.7   Information and knowledge systems

Information relating to PFI/PPP policy should be readily available and communicated to the public sector, private sector participants, users and other stakeholders for raising awareness, and promoting its use. Information should be available on when PFI/PPP procurement should be used, the roles and responsibilities of the public and private sector, processes and stages involved, and decision-making structure. The nature of the information provided is crucial in making informed choices about the opportunities to participate in PFI/PPP projects and to improve clarity and confidence. Good quality of information can therefore have a significant influence on the level of interest and participation in PPP/PFI activities, competition from the private sector, and the subsequent success or failure of projects. It is essential that information on changes and lessons learnt following review and audit of existing PFI/PPP schemes are provided by independent bodies such as the NAO. Other central government support departments such as the Treasury, OGC or public sector advisory agencies (Partnerships UK, 4Ps) also have a key role to play in the dissemination of best practice documents following findings from reviews.

Different channels are used to provide regular updates to stakeholders about PFI/PPP opportunities, status and progress of different projects, their outcome or performance. Information and knowledge management systems include market intelligence reports, seminars, informal discussion with consultants, Private Finance Units (PFU) of government departments, local authority clients, their financial advisers, advertisement in journals and newspapers. Information technology has an increasingly important role to play today in promoting and advertising PFI/PPP opportunities nationally and internationally as well as disseminating knowledge. Some government portals (e.g. Health Information Portal (HIP)) and support agencies provide the platform for disseminating information on best practices, successes and lessons learnt in PPP/PFI projects. For example, the Department of Health Estates and Facilities Knowledge and Information Portal share useful information such as Health Building Notes (HBN) to advice project teams on design and planning buildings whilst the Health Technical Memorandum (HTM) provides advice on design and installation of building services. In addition, there are specialised websites on privately financed infrastructure projects including online directory of consultants and advisers for international PPP/PFI projects. Non-IT systems include policy briefing sessions, seminars, newsletters and press releases. Access to information and knowledge helps to reduce transaction costs and create the level of competition required to achieve value for money in PFI/PPP projects. Timely, reliable and relevant information and knowledge systems promote opportunities, generate private sector interest and build market confidence. Feedback mechanisms from earlier PFI/PPP projects are important to develop knowledge and the dissemination of best practices for continuously improving PFI/PPP projects.

2.3   Governing Principles of PFI Projects

The Private Finance Initiative (PFI) is a specific model of PPP which is a service contract between a public sector and the private sector. Kerr (1998) noted that the 'principle underlying the PFI is that, while the government may need to be responsible for the delivery of a particular service, there are advantages to be gained if the private sector assumes responsibility for managing the service and undertaking the investment'. There are several governing principles that underpin PFI policies and its delivery which are outlined below.

2.3.1   Value for money and risk transfer

Value for money is central to the PFI/PPP debate. In the United Kingdom, PFI should be used only where it is demonstrated to provide value for money compared to the traditional public sector funded route. Value for money is 'the optimum combination of whole life cost (capital and operating costs) and quality of services to meet the requirement of the public sector' (HM Treasury, 2004). Public sector bodies put forward a 'value for money' case for procuring a project through the PFI route which rests upon risk transfer and efficiency in service delivery. Akintoye *et al.* (2003) suggested that 'best value' should be assessed in conjunction with other project aspects such as process costs, risk transfer, service quality and wider policy objectives. The PFI option must therefore be compared with the conventional option, which should include a realistic pricing of all services and the value of the risks. The PFI option is selected only if the whole life cost of the private sector bid is lower than the hypothetical risk adjusted Public Sector Comparator (PSC) based on the same level or quality of services.

2.3.2   Whole life cycle commitment

PFI addresses the shortcomings of the traditional procurement by encouraging long-term cooperation and whole life commitment. Projects that are long-term contracts provide the opportunity for both the private contractor and the public sector to consider costs over the whole life cycle of an asset. In traditional procurement, design, construction and maintenance/operational stages are separated. The traditional approaches have resulted in what one commentator referred to as 'Build and Disappear (BAD)' practice (Winch, 2000). It is this lack of whole life commitment and fragmentation which is often criticised for creating conflicts, confrontation and costly buildings that deliver poor services. In PFI procurement, the SPV/SPC or PFI contractor has to 'Build, Evaluate, Stay Throughout' (BEST) the concession period. PFI therefore represents a move from BAD to BEST practice procurement, at least in theory. The whole life approach in PFI leads to efficiencies through synergies between design, construction of the asset and its later operation and maintenance. The outcome should result in a reduction in costs, both for the private contractor and the public sector client, due to innovation and better integration (ACCA, 2002). Davies (2006) argued that by internalising 'project maintenance costs post-construction, there is an incentive to install more efficient types of technology and deliver the project at a lower cost'. The policy outcomes and the benefits in terms of whole life performance of infrastructure facilities and the delivery of public services can therefore be significant.

2.3.3   Facilitating the delivery of 'core' public services

PFI is driven by public sector needs, and the role of the private sector is to facilitate the delivery of core public services. Delivering services requires three major components: physical, personal and institutional infrastructure (Figure 2.2). Physical infrastructure comprises the structures and networks including power supply, water, sewerage and telecommunication systems, and so on. Institutional infrastructure relates to rules that govern an organisation such as the model of health care delivery or financing system. Personal infrastructure refers to the stock of knowledge and skills in the organisation. PPP/PFI is not the same as privatisation as some elements of the physical, institutional and personal infrastructure are always retained or controlled by the public sector. For example, in prisons PFI projects, whilst the security personnel are directly employed by private sector firms, the care and control regime for prisoners are largely influenced, determined or controlled by public sector authorities. Understanding the nature of the relationship between these three components is crucial in developing a strategic framework for the delivery of services. In health and education PFI projects, doctors, nurses and teachers delivering core medical, surgical, nursing and educational services as well as the institutional infrastructure (clinical service model) remain with the public sector. However, the physical infrastructure, that is how the hospital is designed and constructed, and the investment for the assets as well as its operation and maintenance are the responsibility of the private sector.

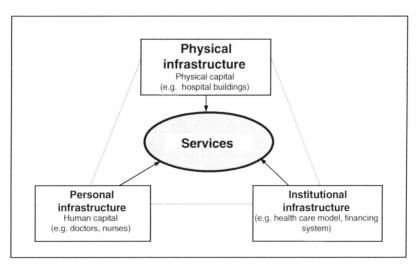

Figure 2.2   Interdependencies of infrastructure system.

Whilst the core staff of nurses, doctors and various medical specialists are retained/employed by the public sector, others mainly maintenance, cleaning, porters and security staff are usually transferred to the private sector/PFI consortium, FM company. However, this transfer of employment or what is often refer to as TUPE (Transfer of Undertakings Protection of Employment) regulation has been the subject of intense debate and a major issue for trade unions in PFI schemes. The public sector retains responsibility for the institutional aspects relating to the health care delivery model and the financing system. The private sector provides support services through integrated hard (e.g. maintenance, groundwork, landscaping) and soft (e.g. cleaning, security, portering) FM for the functioning of the hospital. In other PFI projects such as prisons, FM services include keeping prisoners in custody, maintaining order, control, discipline and a safe environment, providing positive regimes by the provision of education and counselling services and preparing prisoners for their return to the community through pre-release courses (NAO, 2003).

2.3.4   Payments for services based on performance

A key principle in PFI is the link between performance and payments to the private sector based on the successful supply of services to the public sector (Grout, 1997). The payment mechanism provides the incentives for the contractor to deliver exactly the service required in the manner that provides 'value for money' (HM Treasury, 2004). Payments are therefore not received until the asset is fully operational and delivering services. Gruneberg *et al.* (2007) argued that 'if a supplier has a responsibility for how something performs, then his or her contractual liability must extend into the performance period'. This increases operational risks relating to the unavailability of facilities, failure to produce the services required. For example, if the maintenance cost of a hospital turns out to be higher than expected, the PFI contractor has to bear the burden. Certain elements of contract payment are therefore at risk as the link between quality of services and payments provides a powerful incentive for PFI contractors to deliver the standard of services required by the public sector client. The unitary charges or payments made throughout the contract or concession period therefore reflect the performance of the PFI contractor.

2.4   Management Strategy

PPP/PFI projects are complex by nature, so an appropriate management structure is vital for successful implementation. The management structure should reflect the diversity of teams and professionals involved, type of agreements and relationships between the different participants. It is important that the management structure and various agreements address the concern, needs and responsibility of key stakeholders particularly those directly involved or indirectly affected by the PPP/PFI scheme.

2.4.1   Team composition

The SPV/SPC is a single purpose company formed to tender for a specific PFI/PPP project, and if successful will enter into a contract with the public sector client as the awarding authority. The structure and relationships between the stakeholders will vary depending on the type of PFI/PPP scheme. However, the key players would include the client, designers, constructors, facilities managers and financiers who are part of the SPV/SPC. FM contractor plays a key role in the management of the assets delivering hard and soft services as some elements of payments are directly linked to the service performance. Soft FM services are people intensive and often involve services such as portering, security, linen cleaning and catering, and so on. Hard FM services involve estate and building maintenance, associated gardens and ground maintenance to ensure that the building and facilities performs to the required standards. In practice, the FM services are usually delivered by a number of different subcontractors. For example, there could be subcontractors for car parking, pest control, waste disposal, reception, postal and courier services, telecoms, medical equipment, non-emergency patient transport, and so on. Figure 2.3 shows a typical set-up of a PFI/PPP contract.

The public sector client and private sector consortium (SPV) need to have a full range of skills to complete a PFI/PPP contract. Legal, technical and financial advisers are appointed by the public sector to help define business requirements, develop the business case, deal with risk transfer, cost and affordability, payment stream, managing the procurement process and to negotiate the best contract for the client. Appointment of advisers for the SPV/SPC, where there are knowledge gaps, is also important for success at bidding and subsequent stages of the project. As the SPV/SPC is a 'shell' company, its liabilities and obligations to the public sector client are matched by the liabilities and obligations of the subcontractors. PFI transactions are often seen as a three-way relationship between the public sector client, private sector PFI contractor and lenders who provide debt capital and want to safeguard their investment by requiring the PFI contractor to maintain a certain level of debt service coverage ratio.

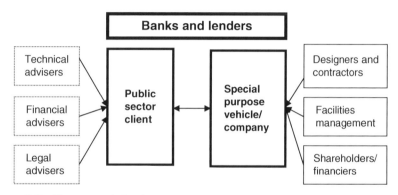

Figure 2.3   Management structure.

2.4.2   Contract and interface management

Depending on the project, the parties involved in the management of the contract will vary. For example, in housing PFI projects, this will include a SPV/SPC, lender, contractor, local authority client, Registered Social Landlord (Housing Association) and an FM provider. The contract will stipulate the obligations of the PFI provider for the construction of new facilities and/or refurbishment of existing facilities. There will be a series of agreements and subcontracts between different parties as shown in Figure 2.4. Over the first part of the contract, usually 2–5 years depending on the size of the project, the SPC will complete the construction and/or refurbishment of existing facilities as required. During the operation and service delivery phase of the contract, the operator/subcontractor will provide certain FM services (both soft and hard FM) as defined in the output specification. For example in housing PFI, services provided will include resident consultation, repairs and maintenance, rent collection, void and tenancy management, waiting list management, caretaking and security (Department of Communities and Local Government, 2008). The scope of services provided is defined by the *'output specification'* setting out the details of the accommodation and FM services the private operator or PPP/PFI contractor is expected to deliver (McDowall, 1999). There will be other parties and agreements involved (not shown in the diagram) such as with insurance companies, the independent certifier and various FM subcontractors working with the FM partner.

The nature of the contractual relationships is fundamental for the successful delivery of complex projects such as PFI/PPP. Smyth and Edkins (2007)

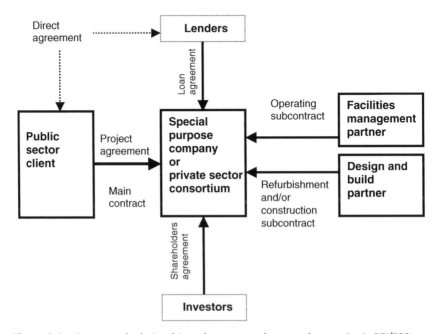

Figure 2.4   Contractual relationship and agreement between key parties in PFI/PPP.

identified trust as a crucial ingredient in the relationship between key actors in PFI/PPP projects. They noted that 'once an SPV/SPC or PFI contractor is appointed switching suppliers is highly constrained legally' and expensive. They also noted that SPV/SPC are 'reacting to structural change in the market represented by PFI/PPP procurement by adjusting behaviour accordingly'. It is suggested that there is need for a shift to relationship management principles focusing on the client interface to manage the relations between the public sector client and the SPC (Pryke and Smyth, 2006).

2.4.3   Stakeholder engagement

A central issue in PFI/PPP implementation is the determination of the impact on stakeholders directly involved or indirectly affected by the delivery of PFI/PPP schemes. Developing an appropriate PFI/PPP implementation strategy requires considerable consultation between the public sector, PFI/PPP contractors and other stakeholders to understand the potential benefits, expectations and risks of different parties. Dowdeswell and Heasman (2004) noted that 'few PPP projects reviewed have achieved real success in involving the ultimate stakeholders – the public'.

The concept of stakeholder analysis is best illustrated with an example in the health sector as shown in Table 2.3. The key stakeholders are identified; potential benefits and expectations and the risks they face are also outlined. In the United Kingdom, a major problem for the NHS Trusts is their inability to meet the increasing demand for health care services due to lack of physical infrastructure to speed up delivery of health services and expand coverage. As part of PFI/PPP framework, NHS Trusts are therefore encouraged to address infrastructure problems through privately financed, design, build and operate hospital projects. The new PPP/PFI projects are expected to generate additional capacity and modern health care facilities to meet the growing demand and to improve the quality of clinical service delivery.

The starting point is to identify the key stakeholders for the project with the National Health Service (NHS) Trust as the public client. This includes the Department of Health or the Treasury who will be involved in assessing whether the benefits sought at the *project initiation or proposal stage* as outlined in the business case provides value for money and there is genuine risk transfer. The key benefits outlined for the NHS client could include several elements: (1) improved physical infrastructure to facilitate health care delivery for patients/consumers; (2) increased capacity to perform more services by treating and admitting more patients; (3) reduce disruptions to nursing, surgical and medical services due to service failures associated with poor performing facilities/buildings; (4) reduce high cost associated with the maintenance of old buildings. The major benefit will be value-for-money argument achieved if the whole life cost of the PPP/PFI project is lower when compared to the traditional procurement approach. Other stakeholders will have different factors to consider. From the users' or patients' perspective, health care services will remain free as this is a fundamental principle of the NHS but there may be concern about introducing or increasing charges for

Table 2.3   Key stakeholders, benefits and risks.

Stakeholders	Potential benefits and expectations	Most likely risks
NHS Trusts	Improve health care delivery	Introduce some service charges (e.g. car park charges affecting low income) Project failure, insufficient demand Patient/consumer or union opposition
Government – Treasury Department	Reduction in expenditure for capital projects Taxes (tax revenue)	Macroeconomic risk (e.g. inflation, interest rates, tax revenue, spending)
Investors	Adequate returns for equity investment	Political and investment risk Not getting adequate returns
Lenders/banks	Debt repayment (interest and principal)	Commercial credit risks
SPC (private developers)	Profit	Political – change of law/government/termination Revenue or demand risk – failure to recover costs of investment due to service failures or lack of demand for facilities Construction risk – delay and cost escalation, liquidated damages Operational risks – failure to maintain and operate facility properly, penalties
Consumers (e.g. patients, patient's relatives)	Reliable health care service	High charges for some services (e.g. car parking, meals, telephones, etc.) Future price increase
Trade unions	Protect employment of members (e.g. public sector employees transferred to private sector)	Loss of jobs Loss of influence and reduction in members contribution/revenue base
Press/public	Clear understanding of project scope and objectives	Concern about future expenditure, payment of higher charges and taxes indirectly by the public
Others, for example Planners Regulatory Authorities	Safeguard consumer interest – quality and fair pricing	Political risk – interference in decision-making process

non-essential services such as car parks, leisure and TV facilities, telephone or food. The benefits or expectations from the private developers, the SPV/SPC are the unitary charges or payments received and profits created from long-term income stream from the public sector client for design, construction, and FM services.

Commercial lenders will, on the one hand, be concerned about whether debt repayment and debt service ratios can be achieved. Investors, on the other hand, will be interested or expected to achieve higher returns on their investment based on internal rate of returns (IRR) and net present value (NPV) calculations to reflect their risks. Government's priority (NHS Trust client, Department of Health, Treasury) may be to reduce or eliminate service charges to the consumers, and to reduce the level of unitary charges/payments to the SPV/SPC so that the project is affordable given the public client's other spending commitments. Other stakeholders such as the press are vital for disseminating appropriate information to the public to ensure that there is a clear understanding of the project scope and objectives, progress, problems associated with the project and to develop public support. In the early PFI projects in the United Kingdom, there were problems relating to how the relationship with the public is managed given the vested interest of trade unions on issues relating to transfer of employment from the public to the private sector. Regulatory authorities such as planners also play a crucial role in safeguarding public interest in PFI/PPP projects by ensuring that planning and environmental issues are addressed. Different stakeholders are involved and each of the parties will view the project risks and benefits from different perspectives. However, decisions on PPP/PFI projects are complex as it is not generally based on the perspective of any one stakeholder. An appropriate balance is required taking into account the benefits and expectations of other key stakeholders and the risks involved. There is therefore a need for stakeholder analysis before the implementation *of PFI/PPP projects*. The stakeholders, their benefits and expectations, and risk profile will influence the financial modelling and funding strategy of a PFI project discussed in Chapter 3.

2.5   Funding Strategy

PPP/PPP projects are funded usually based on the principles of project finance. The source of finance affects the project cost, revenues, risk allocation and therefore the project viability. Sources of finance, whether debt or equity, affects the level of risks, returns, lending terms and conditions such as repayment period, interest rates/charges, foreign currency requirements, project structuring, bankability, the need for various types of guarantees and credit enhancement.

2.5.1   Debt and equity component

Investment for PFI/PPP projects is made up of debt and equity components. Debt capital is provided by lenders (e.g. banks, financial institutions). Equity is usually provided from a variety of sources such as investors, project participants such as design and build and FM subcontractors (see Figure 2.5). The lenders and public sector clients prefer members of the SPV/SPC or SPC to have an equity stake to provide an incentive and long-term commitment to the project. Lenders would normally request an equity typically in the range of 10–30% for non-recourse project financing with debt between 70% and 90%. The debt/equity ratio has implications for the project economics as debt is generally cheaper than equity. High level of equity reduces the payment obligations to lenders (i.e. principal and interest payments) or debt service burden on the cash flow which is crucial at early stages. Due to the

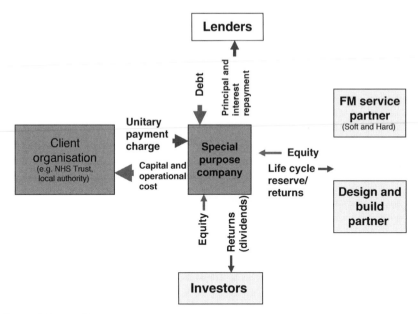

Figure 2.5   Funding structure.

significant contribution from lenders, they will insist on high quality and viability of the project in terms of certainty of revenue income, additional income potential (third party income), established reputation and track record of the companies forming the SPV/SPC and their advisers.

The debt capital and equity are used to fund the project capital and operational cost. In return, the SPV/SPC receives a regular payment/unitary charge to repay lenders interest and principal payments on the debt, returns to investors and shareholders in the form of dividends and to build up 'life cycle' reserves or reserve account for future maintenance and protection of the project.

Debt provided by lenders has the lowest risk, so payments for the principal and interest of the loan have a higher priority; hence, the term senior debt is sometimes used. The equity contribution is more at risk if the project goes wrong but returns are higher if successful. Mezzanine finance called quasi-equity, junior debt or subordinated debt is sometimes required. This type of funding as the name suggests shares some characteristics of debt and equity capital and is usually provided by investment banks. Figure 2.6 shows the relationship between risk and return ratio relating to various funding sources.

2.5.2   Project structuring, bankability and credit enhancement

Sometimes lenders are extremely concerned about the nature of the risks and viability of PFI/PPP projects. Under such circumstances. projects are restructured to minimise risk and to improve bankability. Bankability reflects the level of commercial interest, income stream from PFI/PPP projects and the

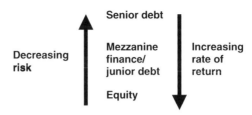

Figure 2.6   Relationship between risk and return based on funding sources.

opportunities for additional revenue streams to meet contractual payments. The key question is often how to structure PFI/PPP projects to attract funding and commercial interest. Lenders sometimes require protection to be built into the concession agreement through credit enhancement techniques as the project company can neither accurately predict nor influence the level of income from usage, demand or price for particular services in future. Credit enhancement strengthens risk management and if appropriately applied can attract debt financing to make PPP/PFI projects bankable.

There are a number of credit enhancement techniques to structure PPP/PFI projects such as minimum volume guarantee which protects future revenue stream. This is useful where there are major problems relating to market risks such as usage and demand. An example is the use of minimum level of usage or traffic volume guarantees in transport PPP projects. Increasing tariff increases the revenue generated which will have a positive impact on the investors return on equity (IRR) and improve the debt service coverage ratio (DSCR). Lenders will normally stipulate the minimum DSCR and investors will expect a level of IRR to be an adequate reflection of the risk associated with a particular type of PPP/PFI project. Increasing tariff and indexation (to protect the value of future cash) is used as a credit enhancement technique to increase level of comfort for lenders by reducing credit risk. 'Tariff indexation' to reflect industry cost increases is also used to deal with key market, demand and pricing risks.

Raising more equity is also another useful technique to reduce debt service payments and increase DSCR but decreases returns on equity. Additional source of revenue or third party income is sometimes used as a credit enhancement technique. For example, in some transport PPP/PFI projects, a developer could have right to build on adjoining land *but* there may be potential conflicts for the SPV/SPC, that is why it is sometimes referred to as a 'single purpose' company. Setting up reserve accounts such as life cycle reserves for unforeseen problems with infrastructure facilities or maintenance can also strengthen a PFI/PPP project as it provides additional comfort to lenders *but* adds to project costs. Setting up escrow accounts where an independent party (agent) appointed to manage revenue account governed by detailed agreement is sometimes used to provide comfort to lenders in international PPP projects where there are significant political, country and macroeconomic risks.

Another credit enhancement technique used in international PPP/PFI projects is the 'partial risk and partial credit guarantees' provided by Multilateral Development Banks (MDB) to protect lenders from unacceptable political and regulatory behaviour. Examples include default activities from host governments such as seizing assets, refusing to pay for services or agreed price increases by the contracting authority. This type of political insurance for lenders has some added advantages such as guaranteeing local commercial bank loans at the earlier (riskiest) stages of PPP/PFI projects (i.e. first 5 years), and to extend the terms of local currency project-backed loans from say 10 to 15 years. Such credit enhancement techniques can facilitate building local capacity for countries at the early stages of PFI/PPP schemes and minimise foreign exchange risk where governments are unable to guarantee foreign exchange payments when tariffs are collected in local currencies.

Basel II Accord requiring the use of more robust credit assessment techniques to minimise risk may significantly affect project lending for PPP/PFI projects in developing economies which may trigger increased support from the MDB. The role of the MDB is expected to intensify in facilitating PPP/PFI projects in developing and middle-income countries. Support for local commercial banks through partial credit guarantee schemes will also become a key part of the MDB strategy to allow PPP projects to be funded locally to minimise exposure to foreign exchange and mitigate political risks.

2.6   Sustainability Strategy

Whole life cycle (WLC), as an economic appraisal tool, is at the heart of PFI/PPP projects due to the long-term nature of such projects (Hosley, 2003). There is a growing recognition that WLC should be linked to social and environmental agenda as part of a client's sustainable development objectives. WLC appraisal can facilitate the understanding of the design and cost implications of sustainability objectives. Figure 2.7 shows how sustainability objectives can be operationalised in PPP/PFI projects.

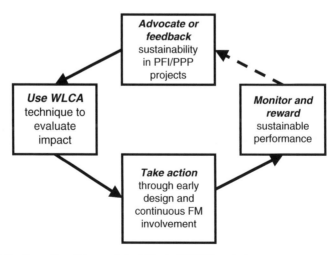

Figure 2.7   Operationalising sustainability in PFI/PPP projects.

Sustainability objectives could relate to reduced material wastage, improved internal environment, increased access to leisure, water and energy efficiency, air quality, safety and security. Using the whole life appraisal technique will enable the selection of the most appropriate design and materials in terms of cost and sustainability (e.g. floor or wall finishes), components (e.g. windows, doors) and systems (e.g. plumbing and heating, hot and cold water, cooling, lighting systems) at the early stages of PPP/PFI projects. Recent conferences on 'whole life costing and sustainability' are a reflection of the growing recognition to strengthen the link between whole life appraisal and sustainability.

There is an increasing awareness that capital cost of buildings or facilities represents only a fraction of the life cycle cost, and the potential to gain efficiency savings comes from the early involvement of the facilities manager. FM therefore has a key role to play in whole life performance and sustainability. The long-term nature and cooperation involved in PPP/PFI procurement means that the public sector client can take a lead in incorporating sustainability objectives and rewarding success. Incentives should be provided to deliver more efficient design solutions to improve the sustainability of buildings. Technical audits should be part of monitoring performance during the service delivery and operational phase.

Fell and John (2003) argued that clients could specify contract clauses and targets or benchmarks linked to payment incentives or penalties. They identified an example of a 50:50 volume risk share on energy use, incorporated in contract clauses during negotiations. Similar approaches could be developed for other environmental performances such as water consumption, wastewater disposal and use of materials. Sustainability objectives relating to social aspects can also be monitored. For example, social objectives such as community engagement, noise, safety, use of local labour, local SME and provision of training can be assessed and benchmarked using performance measures. Bidder's performance on social and environmental issues on previous projects could be reviewed as part of the selection criteria of future projects.

Client bodies recognised the importance of developing sustainability policies covering both social and environmental issues in PFI/PPP projects (Bootland et al., 2003; Fell and John, 2003). For example, the Single Living Accommodation programme by the Ministry of Defence (Defence Estates) was largely driven by quality accommodation and access to leisure facilities. The National Health Service (NHS Estates) has also developed specific tools to address environmental and social performance such as design quality and internal environment.

2.7   European and International Perspective

The market for PFI/PPP is predicted to rise worldwide and there are growing opportunities in Europe, United States, Australia, Canada and other advanced countries, the emerging markets of Central and Eastern Europe and developing countries. Many countries have now implemented

Table 2.4 PPP activity in European countries.

Country	PPP activity
France	Signed PPP projects represent about 2.8% of total number of projects and 3.9% of the value in Europe
Netherlands	Signed PPP projects represent about 1.0% of total number of PPP projects and 1.7% of the value in Europe
Germany	Signed PPP projects represent about 2.4% of total number of projects and 2.9% of the value in Europe
Spain	Total of 92 PPP projects (8.6% of total number of projects) and 12.8% by value in Europe. It has become the second largest PPP market with steady increase in number of projects reaching financial close every year
Greece	Signed PPP represent about 0.6% of total number of projects and 3.9% of the value in Europe. Relatively large investment due to the size of few large PPP projects such as Athens International Airport
Portugal	Signed PPP projects represent about 2.3% of total number of projects and 3.9% of the value in Europe
Ireland	Signed PPP projects represent about 0.7% of total number of projects and 0.7% of the value in Europe
Italy	Signed PPP projects represent about 2.1% of total number of projects and 3.7% of the value in Europe
Hungary	Signed PPP projects represent about 0.8% of total number of projects and 2.7% of the value in Europe

Source: Blanc-Brude *et al.* (2007).

PPP programmes with dedicated PPP units to facilitate implementation. The role of PPP units as knowledge centres for developing policy and strategic framework, implementation and to facilitate capacity building is discussed in Chapter 8.

The United Kingdom is the market leader in Europe accounting for about three-quarters by the number of projects and 58% of the total capital value of European PPP projects (Blanc-Brude *et al.*, 2007). It was noted that PPP appears to have the 'most macroeconomic and systemic significance' in the United Kingdom, Portugal and Spain. But there is growing evidence that PPP is now spreading to continental Europe as a result of recent enabling legislation in countries such as France, Ireland, Germany, Greece and the Czech Republic (Table 2.4). From 1990 to 2006, there were a total of 1066 signed PPP projects in Europe, ranging from 2 (1990) to 152 projects in 2006, with an estimated value of about 200 billion euro (Blanc-Brude *et al.*, 2007).

Six countries – United Kingdom, Spain, France, Germany, Italy and Portugal – account for about 95% of PPP projects (by number) in Europe. However, there are significant differences in the sectoral composition of PPP projects in the United Kingdom compared to other European countries. In the United Kingdom, hospitals are a major part of the PPP programme, reflected by 31% of the number of projects, followed by schools (25%), accommodation type – government buildings, nursing homes, military and prisons – (14%) and the transport sector with a share of 6% (Blanc-Brude *et al.*, 2007). However, in value terms, hospitals make up about 20% of the investment with transport having the largest share of 36% mainly due to the London Underground PPP. Transport projects also dominate the PPP market in Europe reflected in about 60% of the number of PPP projects and 84% by value (Blanc-Brude *et al.*, 2007). The other sectors include hospitals, accommodation, schools, defence and municipal services.

Outside Europe, there are major PPP markets in Australia, United States and Canada with significant investment across different sectors. In developing countries, there are major PPP markets in India, Mexico, Brazil and South Africa. South Africa developed a scheme called the Asset Procurement and Operating Partnership System (APOPS) as a programme within the broader public-private partnership (PPP) framework (Merrifield *et al.*, 2002).

2.8   Concluding Remarks

Successful implementation of PFI/PPP projects clearly depends on developing an appropriate policy and strategic planning framework. The chapter has discussed the essential elements in terms of policy and strategy necessary for the development of PFI/PPP projects to improve the delivery of public services. The environment, institutions, policy theory and objectives, expertise and resources, information and knowledge management systems, processes and monitoring and evaluation mechanisms are identified as central to the policy process. For example, an effective information and knowledge management systems will facilitate learning in both the public sector and private sectors undergoing a major challenge in terms of procurement practices. This should lead to greater knowledge sharing and improvements in the delivery of future PFI/PPP contracts. Evaluation and monitoring is particularly critical to measure and assess policy impact, and more importantly, to ensure that lessons learnt are transferred or fed forward to future projects.

This chapter has also shown that it is necessary to understand how policy elements and strategic framework are linked. For effective implementation, key strategic issues should be addressed so that relevant stakeholders are identified and engaged earlier in the process to develop a viable strategic framework that will reflect the concern of all stakeholders. Managing relationships, developing funding options, structuring projects and dealing with the increasingly complex issues surrounding sustainability are major strategic components which if developed properly will lead to successful implementation. Creating long-term sustainable partnerships and relationships between the public sector and private sectors with mutually agreed goals is crucial to maximise the utility of both parties and to minimise opportunistic behaviour in PPP/PFI programmes. The next chapter explores the actual implementation and delivery mechanisms for PFI/PPP projects.

References

ACCA (Association of Chartered Certified Accountants) (2002) *PFI: Practical Perspectives*. Certified Accountants Educational Trust, London.

Akintoye, A., Hardcastle, C., Beck, M., Chinyio, E., and Asenova, D. (2003) Achieving best value in private finance initiative project procurement. *Construction Management and Economics* 21, 461–470.

Blanc-Brude, F., Goldsmith, H., and Valila, T. (2007) Public private partnership in Europe: an update. European Investment Bank Economic and Financial Report 2007/03.

Bootland, J., Head, P., and Logan, S. (2003) *Sustainability and PFI, CIEF (Construction Industry Environmental Forum?) Meeting Notes 25th November*. Commonwealth Institute, London.

Davies, J. (2006) *Risk Transfer in Private Finance Initiatives (PFIs) – An Economic Analysis, DTI*. Industry Economics and Statistics Directorate (IES) Working Paper.

Department of Communities and Local Government (2008) *The Private Finance Initiative for Housing Revenue Account Housing – The Pathfinder Schemes Baseline Report*. Housing Research Summary, Number 241.

Department of Health (DoH) (1994) *Capital Investment Manual*. HMSO, United Kingdom Official Publications Database, London.

Department of Health (DoH) (2004) *Principles of Generic Economic Model for Outline Business Case Option Appraisal*. HMSO, United Kingdom Official Publications Database, London.

Dowdeswell, B., and Heasman, M. (2004) *Public Private Partnerships in Health: A Comparative Study*. Report Prepared for the Netherlands Board for Hospital Facilities, EU Health Property Network with Centre for Clinical Management Development. University of Durham, Durham.

Fell, D., and John, R. (2003) Sustainability Incentives: Contracts and Payment Mechanisms, CIEF (Construction Industry Environmental Forum?). Meeting Notes 25 November. Commonwealth Institute, London.

Grout, P.A. (1997) Economics of the private finance initiative. *Oxford Review of Economic Policy* 13(4), 53–66.

Gruneberg, S., Hughes, W., and Ancell, D. (2007) Risk under performance-based contracting in the UK construction sector. *Construction Management and Economics* 25, 691–699.

HM Treasury (1997) Appraisal in Evaluation of Central Government, 'The Green Book'. HM Treasury, London.

HM Treasury (2004) *Value for Money Assessment Guidance*. HMSO, London.

HM Treasury (2009) *Safeguarding Government Infrastructure Investment*. Available online at www.hm-treasury.gov.uk (Accessed 14 March 2009).

Hosley, A. (2003) *Life Cycle Costing and Environmental Performance: Making the Link*. CIRIA Members' Report E3142.

Kerr, D. (1998) The private finance initiative and the changing governance of the built environment. *Urban Studies* 35(12), 2277–2301.

Martin, L.L. (2004) Bridging the gap between contract service delivery and public financial management: applying theory to practice. In Khan, A., and Hildreth, W.B. (eds). *Financial Management Theory in the Public Sector*. Greenwood Publishing Group, Portsmouth.

McDowall, E. (1999) Specifying performance for PFI. *Facilities Management*, June, 10–11.

Merrifield, A., Manchidi, T.E., and Allen, S. (2002) The asset procurement and operating partnership system (APOPS) for prisons in South Africa. *International Journal of Project Management* 20, 575–582.

NAO (National Audit Office) (2003) *The Operational Performance of PFI Prisons*. Report of Comptroller and Auditor General, HC 700, Session 2002–2003. The Stationery Office, London.

Pryke, S.D., and Smyth, H.J. (2006) *Management of Complex Projects: Relationship Approach*. Blackwell, Oxford.

Smyth, H., and Edkins, A. (2007) Relationship management in the management of PFI/PPP projects in the UK. *International Journal of Project Management* 25, 232–240.

Wang, L., Yu, C.W., and Wen, F.S. (2007) Economic theory and the application of incentive contracts to procure operating reserves. *Electrical Power Systems Research* 77, 518–526.

Winch, G. (2000) Institutional reform in British construction: partnering and private finance. *Building Research and Information* 28(2), 141–155.

3

Implementation and Delivery Mechanisms

3.1   Introduction

PFI/PPP projects are characterised by three distinct phases of planning and design development, construction and service delivery and operation as shown in Figure 3.1. There are number of distinct activities associated with each phase. By far, the most complicated phases are planning and design development associated with detailed assessment of the public sector or client's needs to justify the project and to choose a preferred bidder and the operation and service delivery phase to ensure that the public sector achieves value for money. From the public sector client perspective, the operation and maintenance phase is the most crucial to ensure that value for money is achieved in delivering services. Each phase is associated with specific steps or stages to achieve the objectives of the PFI project.

This chapter discusses the implementation and delivery mechanisms in PPP/PFI projects. The key stages and issues associated with the delivery of PPP/PFI projects from planning and design development, construction to operation and service delivery are examined. The planning and design development phase examines technical and financial issues such as preparing the business case for the project, invitation and pre-qualification of potential bidders, design solution, evaluation of bids to determine value for money and affordability, selection of the preferred bidder, financial close and developing the full business case for the PPP/PFI project. The construction phase focuses on specific issues relating to completing and translating the design into facilities (as a significant part of the design would have been completed at financial close), resources required for the assembly process, scheduling of key construction activities, phasing of projects and decanting. The operation and service delivery phase focuses on key issues relating to delivering various FM services, performance monitoring to ensure services are delivered in accordance with the output specification, payments to the private sector and deductions for service failures.

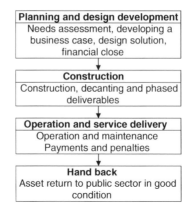

Figure 3.1   Key PPP/PFI phases.

3.2   Needs Assessment and Business Case Development

There are number of activities carried out during the planning and design development phase by the public sector client and the private sector to achieve key deliverables (see Figure 3.2). The process starts with the needs assessment stage where the core objectives of the PFI/PPP project are established by public sector clients and their advisers. There will be a number of options to explore, including the 'do nothing' option. For example, a new build facility may be needed because of shortages of spaces, the need to bring together facilities from different locations to benefit from economies of scale or to replace poorly performing buildings costing too much to maintain. A fully refurbished facility or a minor upgrade may also be required for greater productivity of staff or for more efficient energy utilisation to meet the sustainability objectives of a client organisation. Whatever the type of need, it is critical for the public sector client and their advisers to bring forward project proposals that are well thought out reflecting a genuine business need with strong viability. It is also important that the public sector client and their advisers address issues of risk transfer and develop output specifications that represent a functional solution. The output specification provides the

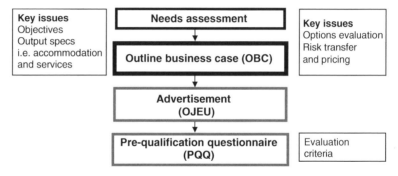

Figure 3.2   Planning stages in PFI/PPP projects.

basis for (1) costing and comparing the traditional procurement with the PFI option, and (2) determining whether value for money is achieved. The output specification is therefore a key document in the planning and design development phase as it sets out the requirements of the public sector client.

At the planning and design development phase, the public sector has a major responsibility to justify the case for a PFI/PPP project and to demonstrate that the PFI solution provides value for money compared to other alternatives including the 'do nothing' and the traditional procurement options. The public sector can and usually relies on technical advisers, legal and financial specialists to assess their needs for a project, develop a strategic outline case and an outline business case (OBC). For example, at the OBC stage, technical advisers such as health planners, epidemiologists, education planners, transport planners, architects, engineers, quantity surveyors (cost specialists), town planners and facilities managers provide advice on various aspects. This includes the scope of a project, the objectives, how it fits into the existing estate master plan and defining the client's need in the output specification. The public sector client will also require financial advisers to construct a comprehensive financial model to explore the relationship between key project variables such as capital expenses (Capex), life cycle costs or operating expenses (Opex), risk allocation, taxes, sources of funding, revenues, payments to lenders and equity stakeholders. This is essential to determine cash flows and demonstrate affordability and value for money.

3.2.1   The output specification

According to McDowall (1999), output specification has changed attitudes to specifying buildings and services by concentrating on aspects of performance which are important to clients. There is a significant debate generated by output-based systems. Unlike a technical specification focusing on '*how*' a facility should be delivered by specifying the dimensions, materials, colour and workmanship, an output specification focuses on '*what*' services are required. One PFI contract manager working for an FM company described the output specification as 'the bible' and was referred to when any disputes arose (Robinson and Scott, 2009). It sets out the operational requirements of the project in terms of accommodation standards and services requirements from hard facilities management (FM) services (e.g. building maintenance, groundwork, landscaping, etc.) to soft FM services (e.g. cleaning, catering, security, etc.).

The accommodation standard relates to the design, environmental performance and physical condition to ensure minimum performance of the building in terms of space and other characteristics within the affordability limits set out in the OBC. Table 3.1 is an example of an output specification for accommodation standard showing the key elements relating to a hospital ward (See Appendix A for more details). The service performance standards reflect the scope and level of requirement for each service category, priority for service delivery, the pass or fail criteria for assessing performance and rectification periods if the service fails. Table 3.2 is an output specification

Table 3.1   An example of output specification (accommodation standard).

Purpose and scope	A 32-bed standard ward accommodation is required, which includes 7 larger bed spaces for Level 2 patients. All beds will be allocated on a speciality basis and arranged in a combination of 4-bed bays and single rooms. The accommodation should be a flexible facility to support
	■ All inpatient services requiring a general level of clinical support ■ Patients needing single organ system monitoring and support ■ Level 2 patients requiring more detailed observation or intervention including those stepping up, or down, from higher levels of care, that is Level 3 (intensive care)
Service trends	Patients in the future are likely to require more complex types of treatments, as the more routine work will be increasingly undertaken by local District General Hospitals. This is likely to lead to a higher proportion of more dependent patients with a longer length of stay than that is currently seen, and therefore clinical areas should be designed to allow for flexible management of patients
Workload activity and facility numbers	13 × 32 bed standard bed wards. Each ward to have 8 single rooms (of which 1 is an isolation room) and 3 are Level 2; 6 × 4 bed bays of which 1 is Level 2
People (maximum volumes)	The maximum volumes based on the number of people (staff, patient and visitors) for each functional area (e.g. reception, bed space, waiting area) will be stated

Area	Patient	Staff	Visitors	Total
Reception	4	3	2	9

Work patterns	24 hours per day, 7 days a week
Access and security	All ward entrances will have proximity card security entry system for staff and the main patient/visitor entrance will have a videophone entry system for all other visitors. The entrance to allow staff access to their rest and change area must not be the main ward entrance. Within the ward, the following rooms will be staff-only access by proximity card security entry: Each ward must have a nominated single main entrance for patient and visitor access only. Patients/visitor access will only gain access when a member of staff operates the video entry door control system. Patients will arrive by foot, or on trolley, chair or bed
Patient and staff flows	See figure A.1.7 in Appendix A for details of the patient and staff flows
External key adjacencies	Establish the relationship between key functional areas to facilitate patient and staff flows. For example, it is essential that the imaging facility is located in a particular area as high volume of ward patients will need to access this service
Key design principles	Establish the design standards and relevant documents to comply with. For example, design guide must be read in conjunction with the following documents: M&E Matrix, Infection Control Document, Clinical Planning Exemplar Text

for service standard showing the key elements relating to car park services (see Appendices B1 and B2).

The output specifications for accommodation and services therefore provides an opportunity for bidders to be flexible, to think about the long-term implications of the service trends, work patterns, activity levels, patients and staff flow and to offer innovative design and FM solutions in PFI projects. Pitt and Collins (2006) argued for output specifications to provide bidders with the opportunity to prioritise the service by defining the client's requirements in terms of level of criticality (relating to the event impacting on the asset) and functionality (relating to the assets importance).

A well-drafted output specification is therefore crucial in the design, construction, operation of PFI projects and the successful delivery of long-term services (4Ps, 2005). Developing an output specification is an extremely challenging process, and the public sector clients and their advisers have the task of specifying a wide range of services in a manner that allows innovation from the private sector but not open to misinterpretation.

3.2.2   Risk transfer

Costing of the output specification and the value of risk transfer is important in determining the bid cost from the private sector perspective and to assess whether it represents value for money from the public sector perspective. Risk is an event leading to a variation from the most likely outcome. All projects are associated with some element of uncertainty and risks. Uncertainty generally reflects an unknown factor that could have a negative or positive effect on a project. A risk is generally known as probabilistic risk as the likelihood on projects can be assessed. The traditional view of risks is negative, often associated with harm, loss or other adverse consequences that would worsen the outcome of a project ('downside' variability). However, some risks could have a positive effect and will improve the outcome of a project ('upside' variability).

It is necessary to investigate the type and level of risks involved in a PFI/PPP project, develop a risk matrix and allocate risks (retain or transfer) to the party best able to manage it effectively, whether public sector client or private sector partner. If both parties bear a certain risk outcome, that is known as a shared risk allocation mechanism. The object is not to transfer all the risks to the private sector but free up the public sector client organisation to concentrate on delivering core services. For example, the focus should be on delivering core nursing and clinical services in hospitals rather than to worry about the hospital facilities, hard estate and soft FM services. Akintoye *et al.* (2003) argued that it is a fundamental requirement that appropriate risks are transferred to the private sector. There are various risks associated with the different phases of planning and design development, construction, operation and service delivery in PFI/PPP projects. Table 3.3 provide an example of risk allocation strategy by the SPV/SPC.

Design and construction risks are retained by SPV/SPC but such risks are transferred to the 'design and build' subcontractor where it is

Table 3.2   Example of FM service output specification for car park services.

Definitions	Contains key terms defined in the service specification. For example, 'car park areas' means all car parks and all other areas designated for parking including on road parking for all types of vehicles including but not limiting cars, bicycles, and so on.
Key objectives	Project Co shall provide a comprehensive car parking service including traffic management across the public sector client site(s). The service shall be operable 24 hours per day 365(6) days per year on a planned and ad hoc basis
Key customers	The key customers for this service are patients, staff, emergency services, visitors, traffic/transport department and service providers/contractors
Scope and service requirements	Project Co shall provide the following services and elements, as part of the car parking service so as to meet the service standards: (1) traffic management, (2) car park areas, (3) designated/priority parking, (3) car park maintenance, (4) car park management and administration and (5) security. Each service requirement/element is further explained; for example, car park management and administration include revenue collection and accounting, complaint processing and permit system. Project Co shall provide the minimum requirement for car parking service 24 hours a day 365(6) days per year on a planned and reactive basis as defined in the response and rectification times (see example below)
Response and rectification times	For the purpose of determining response times and rectification times, the failure or request for service shall be categorised as emergency, urgent or routine. For example, emergency means 'events felt to be life threatening or serious enough to cause significant harm or damage'. Routine means 'faults that are not seen as immediately detrimental and not causing significant operational problems'

Category	Maximum response time	Maximum rectification time
Emergency	Within 5 minutes	15 minutes
Urgent	30 minutes	15 minutes
Routine	1 hour	30 minutes

Performance parameters	Performance for each service element is defined in SF type, category and measured based on response and rectification times, stating the measurement period and monitoring method. For example, service elements could be as follows:

1. All 'No Parking' or restricted parking areas are to be kept free of unauthorised vehicles or other obstructions
2. A system of regular inspections is operable and all faults are recorded with the help desk promptly in the agreed manner
3. Adequate permit tracing and tracking facilities are in operation, with appropriate action taken in the event of vehicles displaying incorrect/out of date permits
4. Controls are in place to ensure that internal roadways are kept clear at all times

Key performance indicators	Key performance indicators (KPI) established for each service element and performance score is recorded based on a 'traffic system'

KPI reference	KPI measure	Performance range/score		
		Green	Amber	Red
KO1	No of complaints per month			

Table 3.3 Example of a risk allocation strategy by SPV/SPC.

Retained risks by SPV/SPC	Transferred to public sector	Transferred to FM subcontractor	Shared by SPV/SPC and public sector
Design (e.g. failure to design to brief)	Political	Maintenance/life cycle costs	Force majeure
Construction (e.g. cost overrun or failure to build to brief)	Occupant/tenant-related damage (e.g. housing)	Operation and performance of facilities	Inflation
Stock condition relating to existing assets	Site availability	Innovation and technological risks	Interest rates
Credit	Volume/demand (e.g. changes in demand for patient services)	Non- or poor performance of services	Changes in legislation

subcontracted to a separate firm. Operational risks relating to escalating life cycle costs, innovation and technological changes are transferred to FM companies/subcontractors. If the maintenance and life cycle costs of a hospital turns out to be higher than expected, the FM subcontractor bears the burden. Some operational risks such as power outage, water supply problems and infection control are more aligned to the estate services (hard FM side) which are different from the risks on the soft FM side. Political risks are transferred to the public sector client. The public sector should retain political and occupant risks as they have control on these risks. Other risks relating to force majeure and changes in legislation should or could be shared.

Grout (1997) reported evidence that volume risk is often borne by the public sector but argued that usage is dependent upon quality of assets and associated risks ought to be borne by the builder or SPV/SPC. The private sector attempts to reduce exposure to volume risk such as the demand for their facilities by transferring to the public sector. For example, in prisons PFI/PPP projects, the private sector is often unwilling to take on demand risk because of changes in sentencing policy which the public sector can influence. In the education sector, there is a risk of falling school enrolment as a result of changes in population parameters (Ball *et al.*, 2000). Asenova and Beck (2008) cited an example of a housing PFI project where the public sector had to consent to the lenders/banks refusal to accept the transfer of volume risk to the SPV/SPC.

In general, public sector clients have to demonstrate sufficient risk transfer to achieve off-balance sheet treatment available for PFI projects (Asenova and Beck, 2008). Volume risk if transferred to the private sector, availability of facilities, non- or poor performance of services, maintenance, life cycle, innovation and technology risks directly affect the payment received or revenue of the private sector operator.

3.2.3   Risk pricing

Problems have occurred in conventionally procured projects because of failure to identify potential risks and to value and manage them. This is usually

Table 3.4 Valuation of risk in PFI/PPP projects.

Scenario	Probability of event (A)	Cost of event (B) (£ million)	Value of risk (C) = (A) × (B) (£ million)
Project completed below budget by £10 m	0.10	−10	−1.0
Project completed on budget	0.20	0	0.0
Project overrun by £20 million	0.40	+20	+8.0
Project overrun by £30 million	0.20	+30	+6.0
Project overrun by £40 million	0.10	+40	+4.0
Risk adjustment to project cost			**+17.0**

referred to as 'optimism bias' associated with underestimating risks particularly cost and time overruns due to a culture of predicting lowest cost and earliest completion. Mott MacDonald (2002) concluded that the poor performance of large public sector projects in the United Kingdom were rooted to the planning and design team's optimism with respect to risk during project appraisal. As a result, optimism bias, as a technique to take account of certain risks, was introduced by the Treasury for all large public sector projects.

The output specification and risk allocation provides the basis for preparing the Public Sector Comparator (PSC). The cost of the PFI solution is based on the accommodation and service standards but the cost of risk is an important element to be added. Risk transfer must therefore be demonstrated through pricing of risks in the planning process. The public sector must take a realistic view of risk allocation. Where risks are transferred to the private sector, then a reasonable price adjustment is expected to reflect the risk transfer strategy. The PFI process involves exploring risk allocation and the value of risk transfer to establish whether the PFI option provides value for money when compared to the traditional route.

The value of risk is quantified based on the probability and the monetary impact should an event occur. Table 3.4 shows a simple example of how risk of cost overrun is valued based on the probability of the event happening, and the cost of the event. The project estimated construction cost is £200 million and the likelihood of cost overrun reflecting various risks is shown to amount to £17 million. Similarly, the value of risk for every other capital and operating cost element is determined. Risk measurements are based on statistical data but capping or reallocation of risks is sometimes used to limit exposure and the impact on project cost.

Risk is valued based on (1) the probability of the event occurring and (2) the costs should the event occur. For each risk event, the process is repeated to arrive at an estimate of the cost or financial consequences. Costing of risks is therefore crucial in PFI/PPP projects as the financial consequences will play a key role in determining affordability and value for money during contract negotiation. Typically, PFI projects seemed to value risk transfer at around 30–35% of construction costs (ACCA, 2004). Pollock and Vickers (2002) highlighted a case where the cost of a PFI hospital became lower than the publicly funded hospital only after including risk transfer. In other words, the 'value-for-money' case rested upon risk transfer at the design, construction and operational stages. Value for money is achieved through the transfer of risk to the private sector. Hence, the concept of risk and its valuation is important in the demonstration of value for money.

3.3   Advertisement, Pre-Qualification and Bidding

It is mandatory to advertise PPP/PFI opportunities in the Official Journal of the European Union (OJEU) as a member of the EU. The advertisement stage involves putting out a set of bid documents for prospective bidders about the proposed project – its scope, objectives, and so on. Bidders are shortlisted using a pre-qualification questionnaire (PQQ) based on a number of technical and financial criteria to identify contractors or teams with the experience and financial standing to successfully deliver PFI projects. The PPQ shown in Table 3.5 is to enable a thorough evaluation and to choose potential bidders who have the capacity, capability and financial resources to undertake the project.

Once shortlisted based on fulfilling the evaluation criteria and achieving a particular overall score (see Table 3.6 for an example of evaluation matrix), the bidders respond to the tender documents which normally include several volumes of output specification, operational policies and standards, payment and performance system and contract agreements.

The bidding involves interpreting a number of documents which usually include an overview of the project, different volumes of output specification to reflect the range of accommodation and FM service requirements, design guide and standards as well as the project agreement. The private sector consortium is expected to address specific issues relating to contract agreement, design, construction, operation and maintenance and project management. The preparation of the bid can be costly, because of the high transaction costs due to information requirements, significant design element and lengthy negotiation period requiring upfront resources. Private contractors bidding for PFI projects incur a higher cost for developing design solutions using output specifications, negotiating contract terms and funding over a long-term period. The costs associated with bidding and negotiation can be disproportionately high for smaller PFI projects. A major criticism therefore relates to the high cost of organising bids which prevents participation particularly from smaller organisations. Recent work carried out by the Audit Commission suggests that bidding costs of the private and public sector together amount to between 5% and 15% of the capital cost, with an average of around 7% in the education sector. Whilst an element of competitive pressure on bidders is necessary to achieve value for money, it is important for public sector or governments to have serious consideration on the size of the schemes and number of bidders invited at key stages to reduce bidding and transaction costs.

Table 3.5 Pre-qualification criteria and checklist.

PQQ criteria	Description of criteria
A1	Details of the organisation/consortium
A2	1. Type of organisation/status of consortium 2. Certificate of Incorporation (if applicable) 3. Certificate of change of name (if applicable) 4. Evidence of formation of relevant organisation/consortium/shareholding company 5. Detailed information on relevant organisations and their specific roles: design and build contractor and FM service provider (soft and hard FM) 6. Information on advisers (including designers, technical, legal, financial and others)
A3	Contact details for consortium's authorised representative
A4	1. Details of bid manager and other key team members/representative(s) 2. CVs for each key team member 3. Details of capacity of key team members in terms of time allocation and potential time conflicts
A5	Statement in respect of Regulation 14 of the Public Services Contracts Regulations 1993
A6	Details of court actions and/or other legal proceedings (where relevant to the bidder's ability to fulfil the role of finance adviser to the Trust)
A7	Statement on potential conflicts of interest and how these will be dealt with
A8	Evidence of professional liability or indemnity insurance
B1	Previous 3 years of audited financial accounts, cash flow statements, overall turnover and specific turnover for PFI/PPP projects, and other relevant information (e.g. announcement to stock exchange, market)
B2	Experience in raising finance on project finance and PFI-type projects. Name and contact details of bidder's bankers along with confirmation that Trust may contact the bank to obtain a reference, if necessary
B3	Experience of each relevant organisation identified (including third party equity providers) in providing equity on PFI-type projects
C1	1. Details of PFI experience in the particular sector (e.g. health sector) 2. Contact details of 3 client references from projects listed in C1
C2	Details of PFI experience in other sectors (e.g. transport, housing, education, etc.)
C3	Details of non-PFI experience
C4	Details of previous experience where the consortium or relevant organisations in A2(5) have worked together
C5	Details of percentage of staff currently employed in PFI work
C6	Average/total number of employees over the past 3 years
C7	Staff turnover as a percentage of the workforce for the past 3 years
D	Employment and training policies
E	Quality assurance, health and safety and environmental policies

Table 3.6   PQQ evaluation matrix.

PQQ criteria and number		Weighted criteria	Non-weighted criteria	Capability	Capacity	Financial
A	1		Not scored			
	2		Not scored			
	3		Not scored			
	4	15				✓
	5		Pass/fail	✓		
	6		Pass/fail		✓	
	7		Pass/fail	✓		
	8		Pass/fail	✓		
B	1	15				✓
	2	Included in B1				✓
	3	Included in B1				✓
C	1	25		✓		
	2	10		✓		
	3	10		✓		
	4	5		✓		
	5	10			✓	
	6	5			✓	
	7	5			✓	
D	1					
E	1					
		Total weight 100				

3.4   Competitive Negotiation and Dialogue Procedures

Under the competitive negotiation procedure, the invitation to negotiate (ITN) is usually a two-stage process (consisting of preliminary and final stages) with the objective of screening out less attractive bids at the preliminary stage, hence, reducing the number of bidders for final negotiation (see Figure 3.3).

However, a new procurement procedure called competitive dialogue was introduced following an EU Public Sector Procurement Directive. This was subsequently implemented in United Kingdom through Public Contracts Regulations SI 2006/5. The directive became effective from 31 January 2006 'to provide an alternative to the growing use of negotiation on complex projects and to make better use of the private sector's role in delivering innovation' (Rawlinson, 2008). Detailed guidance about competitive dialogue procedure is published by Office of Government Commerce (OGC) and European Commission. The OGC also conducted a 'Lessons Learned Study' based on an Olympic Delivery Authority procurement as part of the process of developing experience in competitive dialogue (OGC, 2007).

The competitive dialogue procedure consists of a series of modified stages in the tendering process as shown in Figure 3.3. There are significant similarities with the negotiated procedure from needs assessment up to the advertisement stage. Rawlinson (2008) noted that the competitive dialogue approach

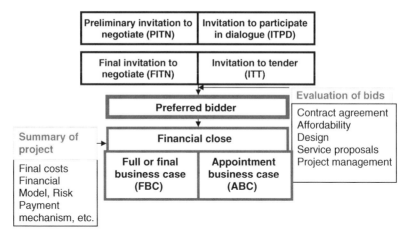

Figure 3.3   Competition stages in PFI/PPP projects.

is expected to be used on 'projects where the client is able to state its requirement at the outset, but either cannot or does not want to define what the solution should be'. The objectives of dialogue are therefore to maintain competitive pressure on all bidders, subject proposals to detail testing and to gradually develop 'compliant and affordable' solutions. Under the competitive dialogue, the public sector tendering option allows for bidders to develop alternative proposals in response to client's requirements. The key difference is that bidders are only invited to submit competitive bids when proposals/solutions are sufficiently detailed. However, Rawlinson (2008) argued that 'despite the substantial difference in process between negotiation and competitive dialogue, the outcome should be similar'. As this is a relative new approach that is largely untested, there are concerns about higher bid costs for public and private sectors, the number of parties and length of dialogue process. It was also noted that 'a typical three-stage dialogue, involving three sets of deliverables and assessments prior to the closure of dialogue could take around 80 weeks, excluding the client's initial development work (Rawlinson, 2008)'. The key differences are summarised in Table 3.7.

3.5   Evaluation of Bids

Evaluation involves a detailed analysis of bids submitted for negotiations or dialogue. The bids submitted are evaluated according to criteria set out in the ITN or invitation to participate (ITP) in dialogue documents. Evaluation focuses on specific areas such as the project agreement, cost and affordability targets, quality of the design solution, the level of FM services and project management structure. Clients and their advisers undertake a comprehensive evaluation to determine compliance and the best bid. Under PFI/PPP projects, there is a greater challenge in evaluating design, cost and the extent to which it meets the output specification. Each bidder has the

Table 3.7   Differences between competitive dialogue and negotiated procedures.

Competitive dialogue	Competitive negotiation
Bidders participate in a first-stage dialogue for developing proposal based on client's output requirements (ITPD stage)	Bidders respond to a fixed set of contractual deliverables based on ITN documents and are required to submit a fully compliant bid (PITN stage). It is also sometimes referred to as invitation to submit outline proposals (ISOP)
Two-way dialogue continues between the client's team and the few bidders invited, with confidentiality respected	Indirect clarification of bidder queries permitted during PITN stage, with all responses copied to all bidders
Proposals are evaluated during dialogue and based on outline solutions and some bidders eliminated at the ITPD stage. During invitation to submit detailed solutions (ISDS), dialogue continues and solutions are tested to confirm compliance with client's requirements. Opportunity to change proposals closed before invitation to tender (ITT) stage	Bidders produce detailed solutions based on output specification and other tender documents. All bids are based on the technical, legal and financial solution described in the ITN documents. Bids with non-compliant solutions at the PITN stage are eliminated and few shortlisted to take forward to the FITN stage
Bids are assessed on the basis of most economically advantageous tender only to select the preferred bidder. No reserved bidder under competitive dialogue	Bids are assessed to select the preferred bidder. A reserved bidder is also identified
Post-bid discussions are limited to clarification and 'fine-tuning' as no major changes are permitted at this stage	Extensive negotiations permitted with preferred bidder with the potential to introduce changes to address problems of design and affordability

freedom to include innovation in design, construction and service delivery. Evaluation team members focus on specific requirements set out in the ITN or ITPD documents. For example, it is usually split between technical, legal and financial, with subgroups to address specialist issues. To manage the evaluation process, clients and their advisers develop an evaluation matrix with performance scores allocated to each criterion. The evaluation is carried out to identify key issues and provide an overall assessment of the attractiveness of each bid. The weighting used to determine the overall score and best bid will reflect the importance of various criteria to the public sector client, particularly if solutions are substantially different which can arise under the negotiated procedure. It is crucial that the bids submitted are compared for affordability and are evaluated for risk transfer to determine whether value for money is achieved.

3.5.1   Value for money

The output specification provides the basis for determining value for money. The PSC is the cost of the project based on the output specification with the public sector as supplier. The PFI bid is the cost with identical output

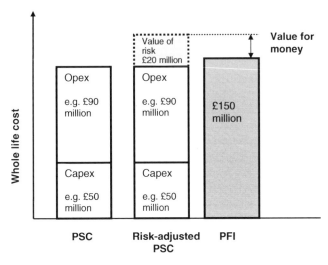

Figure 3.4   Relationship between PSC, PFI solution and VFM.

specification but with private sector as supplier. The cost of the conventional option must include a realistic pricing of all services provided by the PFI solution and must value the risks which are not being transferred. Value for money (VFM) is defined by the Treasury (HM Treasury, 2004) as 'the optimum combination of whole life cost (capital and operating costs) and quality of services to meet the requirement of the public sector'. VFM means that the provision of a service by the private sector results in a net benefit to the public sector. The PFI option is selected if the whole life cost of the PFI bid is lower than the hypothetical *risk-adjusted* PSC. In *net present value* terms, VFM occurs when the privately funded project (PFI option) has a higher net present value (NPV) compared to the PSC. Figure 3.4 shows that VFM is achieved if the whole life cost of the PFI solution is lower than the whole life cost of the risk-adjusted PSC.

The lower whole life cost in the PFI bid from the SPV/SPC is achievable due to competition in the market, the strong incentives to 'reduce costs but not to jeopardize quality' through design and service innovation and better risk management practices from the private sector. Competition is a key factor influencing VFM and is reflected in the level of interest generated from the private sector and the number of bidders. As one PFI Project Director of an NHS Trust puts it, 'insufficient bidders to meet demand are issues for competitiveness'. The Associate Director of a large consulting organisation argued that 'lack of competition drives up prices, causes lack of innovation and cannot demonstrate value for money' (Robinson *et al.*, 2004).

The perceived wisdom dictates that innovation in terms of design and construction leads to operational cost savings (Ball *et al.*, 2000). In order to achieve VFM, it is important to use a competitive tendering process and to maintain competition throughout the process. The Audit Commission (2003) argued that 'if PFI is to deliver value for money to the public sector, the higher costs of private finance and the levels of returns must be outweighed by lower design, construction, management and operating costs'. However, VFM is often the subject of intense debate. VFM assessment should therefore be more transparent. Sussex (2003) argued that whilst PFI probably leads to more projects being completed on time and better maintained hospitals, it may or may not offer design improvements, lower construction costs but probably does not lead to more cost-effective support services. PFI could be cheaper if private firms make significant efficiency savings through innovation in design, construction and management processes.

VFM is assessed during the planning and design development phase from the OBC to the full business case in accordance with the guidance and procedures set out by the Treasury (HM Treasury, 2004). However, the *theoretical* VFM assessment at the early stages to justify a PFI solution depends on the effectiveness of the performance monitoring regime at the operational phase to ensure that *practical* VFM in service delivery is achieved throughout the life of the asset. A PFI project may provide VFM but will be unaffordable if the output specification is too high. VFM is therefore a necessary condition for a PFI project but it is not sufficient on its own as affordability is critical.

3.5.2   Affordability

Affordability or funding gap is a key issue in PPP/PFI projects and is achieved if the cost of the project over the whole life cycle can be accommodated by the public sector client or department's budget, given its other financial commitments. Research have shown that local authorities or public sector clients fail to correctly predict the resources needed and the 'number of times additional funds have to be sought to close a deal is unacceptable' (Robinson *et al.*, 2004). Figure 3.5 shows that the PFI solution is affordable when the whole life budget is at B. However, budget A creates an affordability

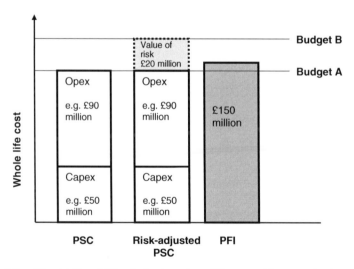

Figure 3.5   Affordability of PFI solution based on different public sector budget.

gap which would require negotiation and further changes in the design and output specification.

Affordability problems are caused by a number of factors. In a recent study on housing PFI pathfinder projects, there were some problems relating to the stock condition survey which affected the level of PFI credits approved to ensure that schemes are affordable to the local authority (Adagun, 2006). The PFI credit is a contribution towards that part of the PFI service charge that is attributed to the capital expenditure. It is intended to cover a portion of the capital and interest costs that would have been incurred if the asset had been procured traditionally rather than through PFI (DTLR, 2002). However, the business plans drawn up by local authority/public sector clients to apply for government PFI credits can creat affordability problems as they sometimes under-estimate the true costs of the capital works due to uncertainty in refurbishment projects. They also do not always reflect the scope of works, the time it would take to reach financial close and therefore made inadequate or no allowance for inflation. Bidders usually treat affordability limit as a target cost but sometimes public sector clients fail to correctly predict the resources needed.

Heavisides and Price (2001) noted that the preparation of output specifications require services far in excess of what is intended can create affordability problems. High bidding costs and lengthy negotiation periods, leading to a reduction in market players have threatened the affordability of schemes in some sectors particularly in health were projects are more complex.

3.6   Preferred Bidder and Financial Close

Following the selection of the preferred bidder, some fine-tuning and adjustments are required before financial close is reached. Under the negotiated procedure, the period to reach agreement on all aspects of the bid can be lengthy and time consuming particularly where there is an affordability gap. Negotiation should reflect the project objectives, criteria set out in the business case and the output specification. However, there are a number of problems encountered such as design issues, risk transfer and value of risk, project coordination and management, cost information and affordability gap. It is therefore important to have a realistic programme for closing PFI/PPP deals. Evidence suggests that the time spent negotiating PFI contracts with preferred bidders (i.e. time from preferred bidder to financial close) is long. For example, Robinson *et al.* (2004) noted that for health PFI projects, the average time taken from preferred bidder to financial close is about 12 months, and 25 months from invitation to financial close. Details about the time scales involved for different sectors can be seen in Chapter 8. According to Audit Commission (2001), the average time taken to complete PFI deals is estimated at 26 and 42 months for local government and NHS projects, respectively. PFI projects are more complex than traditional forms of procurement and therefore require a longer lead-in time before construction commences. The full business case (FBC) or appointment business case (ABC) is then prepared to capture key aspects such as the contract agreement, final whole life costs, the financial model, risk allocation and payment mechanism. The FBC is used by the Treasury to confirm whole life cost, unitary charge, VFM and risk transfer objectives against the hypothetical risk-adjusted PSC. However, due diligence is also required before any contract is signed as all parties will want to ensure that key issues are appropriately addressed. These issues are discussed in the subsequent sections.

3.6.1   Whole life cost plan

Central to PFI/PPP projects is the importance of whole life cost plan to determine the cost required over the life cycle of a project typically for a 20–30-year period (see a simplified example of whole life cost for a power PPP project in Table 3.8).

Whole life cycle appraisal is the systematic consideration of all relevant costs, associated with the acquisition, maintenance and ownership of an asset over its design life (Flanagan and Jewell, 2005). It involves balancing the capital and operating expenses to arrive at a design solution that minimises the total expenditure through the life of a facility. A detailed whole life cost plan is prepared to provide capital and operating cost data for financial modelling and to determine the viability of the project. PPP/PFI projects require resources to pay for the cost involved in planning and design development, construction of the facilities, operation and maintenance, as well as associated project management. However, the cost of resources varies during different phases. Planning costs (sometimes referred to as development costs) are incurred from planning/appraisal, design development up to financial close. Construction cost is incurred from implementation period up to 'commercial operation date' or when the project becomes fully operational. The planning and design development costs and construction costs are often referred to as capital expenditure (Capex). The operation expenditure (Opex) is incurred during the period in which the project operates commercially and produces revenue or income. The different element associated with Capex and Opex as well as other elements for the financial modelling process are shown in Table 3.9.

3.6.2   Financial modelling

The objective of financial modelling is to gain a better understanding of the relationship between costs and revenues, to facilitate decision-making and contract negotiation regarding loan repayments, profitability and return on investment, risk allocation and cost. The analysis should build a clear picture of the interaction between the different elements of cost and revenue covering the life cycle of the PFI/PPP project. It should facilitate negotiations with key stakeholders such as the SPV/SPC, public client, lenders and government. Financial modelling involves developing a financial solution of the PFI/PPP project that reflects the expectation, the key risks and benefits of different parties as demonstrated in the stakeholder engagement and analysis in Chap-

Table 3.8 Example of a simple whole life cost plan for a PPP power project.

Power plant project						Year							
	0	1	2	3	4	5	6	7	8	9	10	11	12
Capital expenditure (in thousands of $)													
Civil works and site preparation	5 000	5 000	5 000										
Plant construction	10 000	15 000	10 000										
Equipment purchase and installation	10 000	10 000	20 000										
Other costs	5 000	3 000	2 000										
Total Capex	30 000	33 000	37 000										
Operating expenditure (in thousands of $)													
Fixed operating and maintenance costs				5 000	5 000	5 000	5 000	5 000	5 000	5 000	5 000	5 000	5 000
Variable operating and maintenance cost				1 000	1 000	1 000	1 000	1 000	1 000	1 000	1 000	1 000	1 000
Fuel cost				20 000	20 000	20 000	20 000	20 000	20 000	20 000	20 000	20 000	20 000
Total Opex				26 000	26 000	26 000	26 000	26 000	26 000	26 000	26 000	26 000	26 000
Total Capex + Opex	**30 000**	**33 000**	**37 000**	**26 000**	**26 000**	**26 000**	**26 000**	**26 000**	**26 000**	**26 000**	**26 000**	**26 000**	**26 000**
Discount rate 10%	1.000	0.909	0.826	0.751	0.683	0.621	0.564	0.513	0.467	0.424	0.386	0.350	0.319
Discounted costs	30 000	29 997	30 562	19 526	17 758	16 146	14 664	13 338	12 142	11 024	10 036	9 100	8 294

Present value of costs/whole life cost
$ 222 587 000
Risk value determined spread across years (not included)
Present value of risk-adjusted cost (not included)

Table 3.9 Examples of data required for financial modelling.

Macroeconomic data	Interest rates Inflation Taxation (e.g. corporate, VAT)
Technical data on project environment (e.g. market/demand)	Type of proposed facilities (e.g. new build, rehabilitation) Existing and projected demand (growth forecast, e.g. school enrolment, health, transport or energy needs, etc.) Proposed location and competition issues
Project cost data	**Capital expenditure (Capex)** Planning, design/supervision costs (professional fees/adviser's fees) Construction costs (including mechanical and electrical costs) Equipment/plant costs **Operating expenditure (Opex)** Operating/admin costs – office, audit, insurance costs; staff costs and wages Life cycle costs/reserves – maintenance costs, cleaning costs, replacement/renewal costs, FM costs
Project revenue data	Unitary charge or payment Fixed charge/availability charge Variable or usage charge Additional/third party income
Financing/project Funding sources	**Debt** Commercial lenders (international, local) Debt/equity leverage acceptable Lending terms and conditions (e.g. payment period, interest rates) Minimum coverage ratio (interest coverage and DSCR (e.g. 1.25), loan life coverage ratio) Currency requirement (hard and local currencies) in developing countries **Subordinated debt (Mezzanine finance)** Lending terms and conditions **Equity** Debt/equity leverage acceptable (e.g. 70:30, 80:20) Return on equity investment (IRR, MIRR)

ter 2. The types of data required for financial modelling are outlined in Table 3.9.

The key costs are the capital expenditure (Capex) and operating expenditure (Opex), some of which are already captured in the whole life cost plan. Other costs include taxes, funding costs such as the interest and principal payments which reflect the balance between debt and equity funding, and the minimum debt service coverage ratio. PPP/PFI projects must have revenues sufficient to cover (1) operations and maintenance costs (e.g. office and staff costs, fuel costs, insurance, etc.), (2) service principal and interest payments on the project debt (DSCR), (3) life cycle reserves where this is required, (4) pay taxes and (5) provide adequate return for equity contributors. The revenue income which determines a project's financial viability depends on key factors relating to the project environment, such as current and projected demand/market data, for example volume of water, output/usage for power projects, traffic volume for transportation projects, occupancy rate for accommodation projects, number of students enrolled in education or patients for hospital projects, and so on. In some cases, additional inputs or natural resources are crucial to generate income from certain PFI/PPP projects such as power stations requiring gas, coal, oil, water, wind, mineral reserves and other raw materials for processing. The costs of inputs or natural resources should therefore be factored in. An example of a financial model for a power PPP project is shown in Table 3.10.

Revenue is determined by tariff or payment structure. In some cases, real or shadow charges are used depending on the type of PPP/PFI project. Real toll charges are generated directly from users and shadow tolls are generated indirectly through payments made by the contracting authority to the project company. Charges are made up of a fixed element/availability charge and a variable element/usage charge depending on performance. For example, in power projects, there is a capacity charge (also known as fixed/availability charge) which is the element of the tariff that is paid even if the plant is not used, to reflect the high capital investment to build the plant.

The usage charge is the variable element depending on level of usage or volume. An indexed element is often included in the revenue or payment formula, usually linked to consumer price index (CPI) or industry price indices (i.e. maintenance, cleaning, etc.) reflecting increases in operating costs or inflation to protect future revenue. There are also penalty charges linked to the revenue or payment formula for non-availability of facilities, low availability or poor quality of services. Revenue could also include third party or additional income designed to strengthened risk management, and to make the project more viable.

Rigorous evaluation techniques are used to assess how well the projects are structured, the risks associated with the projects, particularly demand/market, cost, funding and revenue risks. The key elements of the financial model are the input and output (results) sheets as shown in Table 3.10. The cash flow, balance sheet, and profit and loss accounts are shown as separate modules or components in the financial model. The inputs are the key variables that can be changed such as the demand, various components of Capex (e.g. construction cost, mechanical and electrical costs), Opex (e.g. debt service payments, operation and maintenance costs, corporation tax) and the financing variables (e.g. proportion of senior debt or subordinated/junior debt, equity, interest rates, loan term, etc.) as shown in Table 3.10. The output sheet represents the key results to determine the outcome of the project after choosing various inputs. Depending on the purpose of the financial model and the stakeholders involved, outputs could include key information on rate of return for investors, minimum and average debt service ratio, whole life cost, NPV of usage charge, life cycle charge and the unitary charge.

Sensitivity analysis is used as a technique to assess the impact of changing key input variables on the output/result variables and to identify critical decision variables such as level of unitary or payment charge, income at risk, life cycle costs, level of debt funding required for further examination. In particular, there will be a need to focus on the impact of the cost and revenue changes on the key performance variables such as internal rate of

Table 3.10   Example of financial model for a power PPP project.

(A) Input data and assumptions	Units	Values		
1. Demand assumptions				
Installed capacity	KWh	744 000 000		
Capacity utilisation factor	%	85%		
Output generated	KWh	632 400 000		
Deterioration of capacity	%	2%		
2. Capital expenditure assumptions				
Civil works and site preparation	$	15 000 000		
Plant construction	$	35 000 000		
Equipment purchase and installation	$	40 000 000		
Other start-up costs	$	5 000 000		
Working capital	$	5 000 000		
Total Capex	$	100 000 000		
Annual depreciation rate (20 years straight line)	%/year	5%		
Depreciation expense	$	5 000 000		
3. Financing assumptions				
Debt as a percentage of investment cost	%	70%		
Total debt	$	70 000 000		
Loan term	Years	15		
Interest rate on loan	%/year	10%		
Equity investors required rate of return	%	16%		
Required DSCR		1.3		
4. Operating expenditure assumptions				
Debt service payment (principal and interest)				
Fixed O & M costs (labour, insurance and others)	$	5 000 000		
Fixed O & M costs growth rate (indexation)	%/year	5%		
Fuel costs	$	20 000 000		
Fuel costs growth rate (indexation)	%/year	5%		
Variable O & M costs (5% of fuel costs)	$	1 000 000		
Variable O & M costs growth rate (indexation)	%/year	5%		
Taxes on income	%/year	30%		
5. Tariff/payment				
Usage charge	$/KWh	0.07		
Capacity charge	$/year	40 000 000		
(B) Outputs (results)			Is this acceptable?	
Rate of return (MIRR)	%	17.10%	Yes	
Minimum DSCR	#	1.95	Yes	

Different modules of financial model to explore (modules 0 to 5)

(0) Inputs–outputs	(1) Demand	(2) Capex	(3) Finance	(4) Opex	(5) Cash flow

return (IRR) and debt service coverage ratio (DSCR) required by investors and lenders.

3.6.3   Due diligence

Due diligence is critical for the parties involved in the negotiation of a PFI/PPP project. For lenders, this will involve a number of tasks covering legal and technical issues. First, a thorough evaluation of the proposed project contracts, review of existing legal and regulatory framework, and detailed contract summary reflecting the allocation of risks will be required (legal due diligence). The allocation of project-specific (commercial) risks is also crucial to identify which risks are shared, remain with lenders, project company or government, and more significantly whether they are appropriately covered by contractual arrangements. Given the range of technical and legal issues involved, it is necessary to review the risk allocation matrix to establish whether risks are allocated to the parties best able to manage and control them. Second, detailed technical reviews to ensure that all project information are available, and technical studies (e.g. traffic, health studies or education enrolment forecasts, site and engineering surveys, design), financial and other aspects are carried out by appropriate specialists (technical due diligence). The market condition is one of the most important factors, that is the reliability of the estimates, for example forecast for health or clinical services, transport or water demand, school enrolment, and payment and tariff regime, will have to be examined thoroughly.

The due diligence is usually a structured approach to reflect key issues such as ownership and management structure (if a consortium is involved), reputation risk (i.e. experience of lead firm in project financing), project risk and implementation issues. It is crucial to examine the structure of the project company and to ensure the project is 'bankruptcy remote'. The capital structure is examined, particularly financing issues such as DSCR as this is the primary quantitative measure from a lender's perspective to measure a project's financial strength. DSCR is a ratio of the cash revenue available to repay divided by the principal and interest of the debt. In the early PFI projects in the United Kingdom, the ratio was 1.25 but this has now been reduced to 1.15 reflecting a significant reduction in risk, market confidence and maturity of the PFI/PPP market (Cartlidge, 2006). Similarly, rates of returns or return on equity have generally decreased in line with the reduction in market risks as PFI/PPP market matures. A well-structured project striking an appropriate balance between debt and equity funding is essential to attract investors. The proportion of equity to debt financing depends on the type of PFI/PPP project or sector. Ahluwalia (1997) noted that telecommunication projects with relatively high market risks require a low debt component (debt to equity ratios close to 1:1), whilst power projects with assured power purchase agreements could be funded with debt to equity ratios of 2.5:1 or even 3:1. The reliability of the financial calculations to ensure compliance with tax and other legislation, construction programme, project phasing, project management and quality of supervision, interface management between construction subcontract, and operation and maintenance contract are examined as part of legal and technical due diligence. Macroeconomic data and policies affecting the project such as growth forecast, costs and revenue are also examined as they have significant impact on cash flow and project viability.

3.7   Construction Activities

On appointment, the preferred bidder after financial close will complete the design development and commence construction according to agreed output specifications and performance standards in the FBC. In most cases, a significant part of the design for the PFI project would have been completed. Some design adjustments may be made due to variations or to ensure compliance with regulations, health and safety and standards. This phase involves translating design solutions into a facility or facilities that reflect the requirements in the output specification.

Expert design solution provided by architects, engineers, surveyors and planners with the input of the FM subcontractor is critical to operate and maintain the assets after construction. The construction phase (see Figure 3.6) involves planning for different deliverables (facilities) and decanting, managing and coordinating resource inputs such as labour, materials, components and plant, design teams and subcontractors to complete the PPP/PFI project to the required output specification.

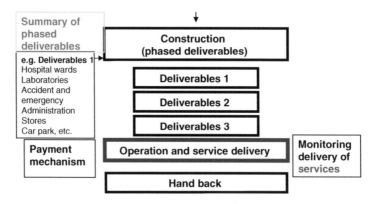

Figure 3.6   Construction and operation stages in PFI/PPP projects.

3.8   Operation and Service Delivery

After construction is completed, the PPP/PFI facility is managed by the concessionaire to deliver the services in the output specification over the full contract period. Inadequate or poor management of service contracts can undermine the public sector client's ability to achieve VFM or reduce the return on investment for the private sector. The principal–agent theory perspective provides a useful framework to understand the relationship between the principal (public sector client) and the agent (SPV/SPC or PFI/PPP contractor) during the operational and service delivery phase of PFI/PPP projects.

As PFI projects are based on the consumption of services, there is a need to have an effective performance mechanism to assess compliance with service level agreements (McDowall, 2000). Firstly, there is a need to understand the output specification in terms of (1) the level of services required and (2) the criticality of various services to the activities of the client. For example, a hospital theatre is more critical than a store or a car park in delivering clinical services. Secondly, there is need to understand the pass/fail standard or scores associated with various services during performance monitoring. Thirdly, it is important to understand appropriate metrics and monitoring methods to determine performance scores and whether a service has passed or failed. The metrics are used to measure the level of services achieved or to determine the performance score (based on the standards set) usually against a percentage scale with a minimum standard and a scale reflecting different penalties if performance falls below a threshold. Finally, there should be a clear understanding of the payment mechanism by establishing a logical link between performance scores and payments received. Provided the accommodation standards and service requirements are met, the service provider will be paid. For risks relating to service performance and non-availability of a facility, penalties are applied to the private sector. However, for this to be effective, penalties in the payment mechanism should be set at an appropriate level and information about service performance and availability should be collected regularly (Ball and King, 2006).

3.8.1   Performance monitoring

Performance monitoring provides a powerful incentive to deliver the standard of services required by the public sector client stipulated in the output specification (Ng and Wong, 2007). Developing a robust performance measurement system (PMS) with relevant metrics to capture a wide range of services and choosing appropriate monitoring methods is therefore a major challenge. Partnerships UK (2006) cited the case of a PFI providing accommodation and training facilities with 361 KPI or performance metrics. Problems can arise from having too many metrics; as a result, a public sector client recently combined the generic and specific monitoring forms to produce around 10 key indicators for each FM service areas. Partnerships UK (2006) also found that a range of methods are used such as formal customer service satisfaction surveys, regular meetings with stakeholders, feedback from the help desk, real-time information systems such as building management systems (BMS) for monitoring performance. For some PFI projects, user satisfaction is particularly relevant to service performance but can create problems where there are multiple layers of users. For example, one hospital manager noted that more complaints were received from hospital staff than from patients, but saw this as positive as faults should be prevented or rectified before they affect end users (Partnerships UK, 2006). Spot checks, site visits and third party audits are also used (McDowall, 2000). However, Ng and Wong (2007) argued that performance 'may not be truly revealed should the frequency of auditing be too sparse' but excessive surveillance would increase project monitoring costs.

Kunz and Pfaff (2002) argued that the function of performance evaluation is twofold: (1) to control discretionary behaviour aligned to incentives and (2) to evaluate contribution to the output and determine compensation based on performance. According to standard agency theory, an optimal incentive contract involves a pay-for-performance scheme linking the agent's pay-off to production indicators (partially) correlated to effort level (Kunz and Pfaff, 2002). In some cases, the agent's 'remuneration is partially contingent on benchmarked performance' (Hensher and Stanley, 2008). However, there are issues related to information asymmetry which gives rise to two fundamental problems. The first is what is usually referred to as 'incomplete contracts', lacking sufficient precision to cover all service delivery contingencies which will create unanticipated service delivery problems. The greater the completeness of contracts, the lower the potential for dispute and abuse. The second problem is agent's opportunism or moral hazard defined by Wang et al. (2007) as 'the risk that the agent can change its behaviour to the detriment of the principal and when a principal cannot fully monitor an agent's actions'. The situation is characterised by the agent choosing actions to relocate his effort to maximise his utility and in doing so will trade off effort costs against the expected monetary (unitary payments) and non-monetary (e.g. prestige or reputation) consequences (Kunz and Pfaff, 2002). However, recent study by Robinson and Scott (2009) found that low level of performance deductions from PFI contractors may not reflect the agent's effort level due to problems associated with interpreting output/performance specification, contract monitoring and PMS. It is therefore important to have a performance-based service contract designed, with a payment mechanism as a critical component, to induce optimal effort and ensure efficient risk transfer.

3.8.2   Payment mechanism

The payment mechanism deals with operational risks relating to the unavailability of facilities, failure to produce the services required in the output specification. The operational risks affect the revenue or the unitary payment received by the private sector operator. There are various payment models for PFI projects, and deductions are made from the unitary charge or payment if facilities are unavailable or there are service failures.

In model A below, the unitary payment is based on available places (e.g. prison, school or hospital places) which include associated core services such as heating, cleaning, mail delivery and food. Although this is a simplified model with few variables, in practice there will be many sub-variables relating to the different types of services, and functional areas and units in a facility. The payment structure is non-separable, so there is a single payment for availability of facility and services as they are included in the definition of an 'available place'.

Example of model A payment structure

$P = (F + I) - Z$

P = unitary payment per place

F = fixed amount per available place per day

I = indexed amount per available place per day (e.g. increased by retail price index)

Z = performance deductions

In model B, the unitary payment is based on the full provision of overall accommodation divided into units and includes associated FM services such as heating, mail delivery and food. This is another example of non-separable or single payment structure and there are separate deductions for unavailability and performance. However, the level of deductions reflects the importance of each unit or type of accommodation if the service provider fails to provide an available place. Deductions are based on the level of criticality of the accommodation or a particular service and the variables used should reflect materiality and proportionality in the operation of the payment mechanism. For example, in a hospital, any shortfall in the standard of basic facilities such as admission wards and operating theatres will have greater consequences on patients receiving health care compared to other non-essential facilities that do not directly contribute to the core clinical activities.

Example of model B payment structure

$P = (F \times I) - (D + E)$

P = unitary payment per place/day

F = price per day for overall accommodation requirement

I = indexation factor

D = deductions for unavailability

E = performance deductions

However, model C is an example of a payment structure that is separable where the unitary payment is divided into separate availability payment stream and FM services payment stream. The availability payment is for the provision of assets such as buildings and equipment, and the service payment is for the provision of FM.

Example of model C payment structure

$P = (A + Q) - (D + E)$

P = unitary payment per unit

A = availability payment

Q = indexed FM payments

D = deductions for unavailability

E = performance deductions

The split between the availability and services payments in model C is crucial in terms of performance risks. Whilst a key principle of PFI is that unitary charge or payment is conditional upon supplying services, which should in theory be reflected in non-separable or single payment structure. In practice, lenders seek to minimise their credit risk through a separate availability payment stream for the capital investment. The extent of the deductions from availability payments is also minimised to comply with the minimum DSCR required by lenders. For this reason, the availability payment is sometimes seen as a fixed charge that changes slightly. Failure to maintain the minimum level of DSCR due to the unavailability of the facility will result in a breach of the loan/credit agreement between the PFI contractor and the lenders who provided the capital.

There are also variations to these models to reflect different issues and concerns of stakeholders. For example, Asenova and Beck (2008) cited a case where the unitary charge was split into three parts – charge for capital repayment, another part was for maintenance over the asset life cycle and another part relating to the equity provided. In other contracts, there may be a variable element or charge that depends on usage, volume or demand factors such as occupancy rate of a hospital ward, or use of sport facilities which is what Handley-Schachler and Gao (2003) refer to as a 'VFM' risks arising from the danger that a service which is very expensive will not be fully utilised. The availability payment usually forms a significant part of the unitary charge which is fixed for the concession period but PFI contract allows for an annual adjustment for inflation and periodic adjustments for the service component of the charge through benchmarking and market testing. Market testing is used for soft services to adjust the payment of services to ensure that VFM principles are followed throughout the operational stage (Boussabaine, 2007). It involves a 're-tender of the soft service provision where the current service provider competes against the best in the market place to earn the right to deliver services for the next 5-year period (Drivers Jonas, 2004).' But it is recognised that market testing can be very expensive and time consuming compared to benchmarking 'where the current service provider can demonstrate that their costs remain competitive'. However, benchmarking is sometimes viewed by stakeholders including FM companies with scepticism and mistrust (Drivers Jonas, 2004).

The payment mechanism depends on the effectiveness of performance monitoring. The 'more difficult a service contract is to monitor, or the effort levels required are high, the greater the transaction costs incurred' (Martin, 2004; Miller, 2005).

3.8.3   Handing back

As PFI contracts specify the condition in which a building is to be handed back to the public sector at the end of the contract, the contractor is incentivised to ensure the building is well maintained throughout the concession period. However, there is the issue of residual value risk which relates to what the net value of the asset will be worth at the end of the contract period. There is the possibility that the public sector client may no longer need the asset at the end of the concession period. A key factor which determines residual value is the asset specificity relating to the alternative uses associated with a particular facility. For example, a private railway company may be able to sell its rolling stock to other companies, but the rail track and its associated infrastructure must stay in place which may limit its value. Certain facilities such as prisons could also have limited alternative uses which may affect their value if it is no longer required (Cartlidge, 2006).

3.9   Concluding Remarks

This chapter has discussed the key phases and stages in the implementation of PFI/PPP projects from planning and design development to operation and service delivery. Key issues and the role of stakeholders at each stage with the key outcome expected are discussed. It is very clear that a successful outcome for PFI/PPP projects not only depends on a robust policy and strategic framework but also requires well-defined implementation processes and steps with unambiguous roles and relationship between stakeholders. PFI/PPP transactions are often seen as a three-way relationship between the public sector client, private sector contractor and lenders who want to safeguard their investment. The implementation processes discussed in this chapter will ensure that the public sector is able to develop a project that fulfils their specific operational needs at the planning stages but viable and attractive enough to generate a level of interest from the private sector, the banks and investors who will provide funding. The competition stages provide a mechanism for evaluating solutions proposed by private contractors to address the needs of the public sector in the output specification and other tender documents. These stages are also associated with particular processes to safeguard the interest of lenders and investors. Key processes are discussed to ensure that the right bidder is selected as inappropriate decisions could have major impact on the level of services delivered during the operational phase. The next chapter examines the specific controls and mechanisms in place to facilitate good decision-making and to ensure that the services delivered are needed, appropriately determined through proper organisational structures and processes.

References

4Ps (2005) *4Ps Review of Operational PFI and PPP Projects*. Available online at www.4ps.gov.uk (Accessed 14 March 2008).

ACCA (Association of Chartered Certified Accountants) (2004) *Evaluating the Operation of PFI in Roads and Hospitals*. Research Report No. 84. Certified Accountants Educational Trust, London.

Adagun, A. (2006) *The Workability of PFI in the UK Social Housing*. Unpublished MSc dissertation, London South Bank University.

Ahluwalia, M.S. (1997) Financing private infrastructure: lessons from India. In: Kohli, H., Mody, A., and Walton, M. (eds). *Choices for Efficient Private Provision of Infrastructure in East Asia*. World Bank, Washington, DC, pp. 85–104.

Akintoye, A., Hardcastle, C., Beck, M., Chinyio, E., and Asenova, D. (2003) Achieving best value in private finance initiative project procurement. *Construction Management and Economics* 21, 461–470.

Asenova, D., and Beck, M. (2008) In: Akintoye, A., and Beck, M. (eds). *Obstacles to Accountability in PFI Projects, Policy, Management and Finance for Public Private Partnerships*. Wiley-Blackwell Publishing, Oxford, pp. 45–63.

Audit Commission (2001) *Building for the Future: The Management of Procurement Under the Private Finance Initiative*. Audit Commission Publications, Wetherby.

Audit Commission (2003) *PFI in Schools: The Quality and Cost of Buildings and Services Provided by Early Private Finance Initiative Schemes*. Audit Commission, London.

Ball, R., Heafey, M., and King, D. (2000) Private finance initiative – a good deal for the public purse or a drain on future generations? *Policy and Politics* 29, 95–108.

Ball, R., and King, D. (2006) The private finance initiative in local government. *Economic Affairs* 26(1), 36–40.

Boussabaine, A. (2007) *Cost Planning of PFI and PPP Building Projects*. Taylor and Francis, Oxford.

Cartlidge, D. (2006) *New Aspects of Quantity Surveying Practice*, 2nd edn. Elsevier Butterworth-Heinemann, Oxford.

Drivers Jonas (2004) *PFI – Soft Services, Lift – Hard Services FM Benchmarking – Are You Prepared?* In Brief, Drivers Jonas property consultants.

DTLR (2002) *Applying for PFI Credits*. The Stationery Office, London.

Flanagan, R., and Jewell, C. (2005) *Whole Life Appraisal for Construction*. Blackwell Science, Oxford.

Grout (1997) Economics of the private finance initiative. *Oxford Review of Economic Policy* 13(4), 53–66.

Handley-Schachler, M., and Gao, S.S. (2003) Can the private finance initiative be used in emerging economies? Lessons from the UK's successes and failures. *Managerial Finance* 29, 36–51.

Heavisides, B., and Price, I. (2001) Input versus output-based performance measurement in the NHS – the current situation. *Facilities* 19(10), 344–356.

Hensher, D.A., and Stanley, J. (2008) Transacting under a performance-based contract: the role of negotiation and competitive tendering. *Transportation Research Part A: Policy and Practice*, 42(9), 1143–1151.

HM Treasury (2004) *Value for Money Assessment Guidance*. HMSO, London.

Kunz, A.H., and Pfaff, D. (2002) Agency theory, performance evaluation and the hypothetical construct of intrinsic motivation. *Accounting, Organisations and Society* 27, 275–295.

Martin, L.L. (2004) Bridging the gap between contract service delivery and public financial management: applying theory to practice. In: Khan, A., and Hildreth, W.B. (eds). *Financial Management Theory in the Public Sector*. Greenwood Publishing Group, Portsmouth.

McDowall, E. (1999) Specifying performance for PFI. *Facilities Management*, June, 10–11.

McDowall, E. (2000) Monitoring PFI contracts. *Facilities Management*, December, 8–9.

Miller, G.J. (2005) Solutions to principal–agent problems in firms. In: Menard, C., and Shirley, M.M. (eds). *Handbook of New Institutional Economics*. Springer, Dordrecht.

Mott MacDonald (2002) *Review of Large Public Procurement in the UK*. HM Treasury, London.

Ng, S.T, and Wong, Y.M.W. (2007) Payment and audit mechanisms for non private-funded PPP-based infrastructure maintenance projects. *Construction Management and Economics* 25, 915–924.

OGC (2007) *Competitive Dialogue Resources*. Available online at http://www.ogc. gov.uk/procurement_policy_and_application_of_eu_rules_specific_application_issues. asp (Accessed 25 August 2009).

Partnerships UK (2006) *Report on Operational PFI Projects.* Available online at www.partnershipsuk.org.uk (Accessed 14 March 2006).

Pitt, M., and Collins, N. (2006) The private finance initiative and value for money. *Journal of Property Investment and Finance* 24(4), 363–373.

Pollock, A., and Vickers, V. (2002) Private finance and value for money in NHS hospitals: a policy in search of a rationale? *British Medical Journal* 324, 1205–1208.

Rawlinson, S. (2008) *Procurement: Competitive Dialogue, Building Magazine.* 23 May, pp. 64–66.

Robinson, H.S., Carrillo, P.M., Anumba, C.J., and Bouchlaghem, N.M. (2004) *Investigating Current Practices, Participation and Opportunities in Private Finance Initiative.* Loughborough University, Leicestershire.

Robinson, H.S., and Scott, J. (2009) Service delivery and performance monitoring in PFI/PPP projects. *Construction Management and Economics* 27(2), 181–197.

Sussex, J. (2003) Public-private partnerships in hospital development: lessons from UK's private finance initiative. *Research in Health Care Financial Management* 8(1), 59–76.

Wang, L., Yu, C.W., and Wen, F.S. (2007) Economic theory and the application of incentive contracts to procure operating reserves. *Electrical Power Systems Research* 77, 518–526.

4

Governance in Project Delivery

4.1   Introduction

Governance relates to the controlling of processes and actions taken by people to deliver the desired 'outcomes' for any project. The term is commonly associated with standards and procedures that are routinely followed with underpinning mechanisms in place to ensure compliance (Williamson, 1996). For public-private partnership (PPP)/PFI projects, the project governance will default to the Vice Chancellor for higher education, or the local education authority (LEA) for school projects as they are responsible for the physical infrastructure while the school and its Board of Governors are responsible for the delivery of education, cost of staff and services (Edwards and Shaoul, 2003). The Chief Executive of an NHS Trust Board is responsible for health PFI/PPP projects (NHS, 1999). Despite this, the National Audit Office (NAO, 2006) found that there was lack of clarity in key accountabilities and roles relating to governance in major PPP/PFI projects. In some sectors, there is an added complexity in governance due to the relationship of the organisations and the interests between the different stakeholders involved. For example, in the schools PFI projects, financial decision-making processes are divided between the LEA and the school's governing body, and this has been shown to affect decision-making and the operation of the control processes (Edwards and Shaoul, 2003).

This chapter starts with some definitions and principles of 'governance'[1] , and using the project development process, discusses and analyses the organisational structures, critical decision-making points and the controls and monitoring mechanisms available for 'governing' the development of a PPP/PFI project from one delivery stage to the next. This chapter also examines and reviews critical success factors for project management during the construction phase and reference will be made to government's attempt to improve public services through better delivery of major capital schemes, including PPP/PFI projects.

---

[1] The focus of governance for this book is in relation to management processes.

4.2   Definitions and Principles of Governance

Governance is a term often used for describing the processes and systems by which an organisation or society operates (Winch, 2001). Frequently, a government or organisational body of people is established to administer these processes and systems to ensure project outcomes are not hindered or compromised. The word 'governance' derives from Latin origins that suggest the notion of 'steering', and hence, many organisations have an oversight or steering board governing the organisation's business developments or projects (Civil Service Governance, 2008). The term 'governance' is also applied generally in industry (particularly in the information technology sector) to describe the processes that need to be in place for a successful project outcome (CCTA, 1996; OGC, 2006).

The Office of Government Commerce (OGC) defines governance as a 'concern with accountability and responsibilities' and describes how an organisation is best directed and controlled through a rigorous and robust governance structure. The published report (OGC, 2006) identifies three key elements of governance. First, for the organisation, this includes organisational units, structure and coordinating mechanisms. Second, in terms of management, the roles and responsibilities established to manage business change, operational services and the scope of the power and authority, which they exercise. The third key element is policies, the frameworks and boundaries established for making decisions about investment in business change (OGC, 2006).

Corkery (1999), in his study of concepts and applications of governance, simplified governance into macro-level covering territories and multiple functions, mezzo-level covering policy sectors, functions or generic organisational types and micro-level covering individual organisations. Winch (2001) focused on the micro-level and developed a conceptual framework for understanding the governance of construction project processes, drawing on transaction cost economics. In the case of PPP/PFI projects, it should be noted that concerns are not limited to micro-level governance as these projects are frequently subject to mezzo-level influences such as government policies. Further support for the need to understand the mezzo-level influence on governance comes from a study by Mustafa (1999). This study demonstrated that policy makers are the dominant influence in determining the development of PPP/PFI projects. In the United Kingdom, this is demonstrated through major changes following government reviews of key processes of PFI/PPP/PFI and the introduction of new structures, standard documentation and tools to improve the governance of PPP/PFI projects. Hence, in any PPP/PFI project, governance at all levels must be of a significantly high standard to ensure and attract private sector participation in the delivery of public services (Badshah, 1998).

4.3   Key Components of Governance

Governance helps to outline and understand the relationships between all internal and external groups, teams or stakeholders involved in a PPP/PFI

Table 4.1   Key components in governance.

Component of governance		Key features and examples
Organisational structure	Roles/ accountability	▪ Organisational and project units ▪ Relationship between organisational and project units ▪ Defined responsibilities for each organisation/project unit
	Project team/project forums	▪ Defined terms of reference for each team/project group ▪ Participants who should contribute ▪ Coordinating mechanism and communication strategy ▪ Authority to make decisions
Control and monitoring mechanisms	Standards	▪ Standard processes (e.g. for pre-qualification, evaluation) ▪ Use of standard documentation (e.g. contract documents) ▪ Reference documents (e.g. best practices) ▪ Policies (e.g. procurement rules, building standards)
	Tools methodologies	▪ Tools to support projects (e.g. risk management, value-for-money guide) ▪ Tools to support operational areas (e.g. project management tool)
	Compliance	▪ Mechanisms for monitoring compliance (e.g. output specification, performance measurement and payment mechanisms) ▪ Collection and analysis of metrics (e.g. benchmarking) ▪ Audits of projects (e.g. post-project evaluation, performance scores)

project. It ensures the proper flow of information or reporting structure to allow informed decision-making by appropriate governing boards. The components of project governance and how they may be applied to PPP/PFI projects are shown in Table 4.1.

The governance in terms of organisational structure together with the control and monitoring mechanisms such as the tools and standards to support the planning/development and implementation of a project will ultimately ensure that the project has a clear direction and is coherent with the objectives of the commissioning organisation, whether it is a Hospital Trust, school, local authority or a highway authority.

4.3.1   Organisational structure

There is some literature identifying project managers as the single most critical factor affecting successful project delivery (Hartman, 2000; Bandow, 2001; Powl and Skitmore, 2005). However, the organisational structure in which they operate is crucially important. In the education context, the LEA is expected to consult the school governing bodies in the development of school PFI projects through an appropriate organisational structure. The school governors are in turn accountable for the decisions that are made to the other stakeholder groups. An effective organisational structure which

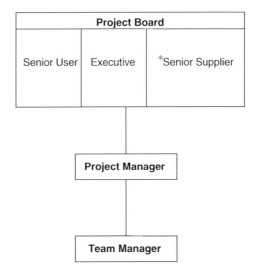

Figure 4.1 Organisational structure showing simple lines of accountability. ∗ denotes *post-selection of preferred bidder/service supplier (SPC/SPV)*. *Source*: CCTA (1996).

defines the reporting structure, lines of accountability, the project manager's role and responsibilities within a project team is therefore critical to ensure and maintain a good governance structure and mechanism for reporting project plans, decision-making, contingencies and delivery status to the overarching Project Board. An example of a simple organisational structure is shown in Figure 4.1.

Within the context of a health service PPP/PFI project, the senior user could be the Medical Director representing all clinicians, the senior supplier[2] is a Director from the project company (SPC/SPV) and the Executives are Directors from the NHS client side. Also on the board are other stakeholders both internal and external to the procuring NHS organisation such as council, government departments, voluntary services, medical school and any other key stakeholder considered to be important to ensure decisions can be made timely and effectively (OGC, 2008). The inclusion of key stakeholders at the right level onto the board is essential for stakeholder buy-in from the outset and ensures that this is sustained throughout the life of the project.

For improved project delivery through effective accountability, the key requirements at the organisational level are for the top or senior management team/Project Board to be clear about strategic goals and the roles and relationship between the different organisations and stakeholders involved. This is essential for overseeing the portfolio of major programmes and managing project risks against project capability (OPSR, 2003). At the programme

---

[2] Senior supplier is co-opted onto the board for development of the full business case.

or project level, there must be an understanding of strategic departmental priorities, identifying and managing risks and interdependencies with regular independent scrutiny of progress (OGC, 2003). A single-named individual, a senior responsible officer (SRO) or equivalent needs to be accountable for each major programme, and at the project level, the working teams need clear roles and responsibilities and a vision translated by the project team into a plan with milestones, regular reporting and review together with stakeholder involvement (DoH, 1994). Stakeholders, internal and external to the organisation, are critical to the success of the project and a stakeholder analysis should be completed during the early stages of project development to ensure key stakeholders are represented and involved at the right decision-making forum in the projects' reporting structure.

4.3.2   Control and monitoring mechanisms

The control and monitoring mechanisms include a range of standards and tools to support effective planning and implementation of projects and to monitor compliance to ensure that the project objectives are achieved.

For example, the perception in United Kingdom has been that construction projects in the public sector have difficulty in delivering on time, budget and within the scope of the project (Latham, 1995; MacDonald, 2002). The government have therefore invested heavily in creating a wealth of control and monitoring tools to ensure compliance and share or transfer knowledge on 'good practice' and 'project governance' through the OGC. For example, a control directive exercised by the OGC Supervisory Board requires that all requests for proposals and invitation to tender for complex projects in government should refer to the 'Best Practice' guide. It is also important to confirm compliance with it at every stage of planning and development phase (OGC, 2002).

There are many tools and standards developed to facilitate control and monitoring during the delivery of projects such as the approval mechanism using prescribed value-for-money accounting and financial methodologies and discounting techniques. For example, the bidding process is designed to create competition and to maintain competitive pressure throughout to ensure that the private sector PFI/PPP bid provides value for money. However, one of the key issues particularly in PFI/PPP projects is risk management. Standard documents and tools have been developed for risk allocation which is continuously evolving based on best practices to capture the different types of risks, and determine which party is best able to manage those risks at key phases of planning and development, construction and service delivery and operation. Project risk varies during the different stages of a contract, but the planning and development phase is one of the most prone to risk due to the complexity of processes involved from the start to bidding, negotiating the contract, awarding the contract and preparing the final business case. Risk management is therefore crucial in PFI/PPP projects. However, most projects do not use or rely heavily on quantitative risk analysis[3] but virtually all projects make use of qualitative risk analysis as a management tool for project control (Cartlidge, 2004). Qualitative techniques identify, describe and assess risk and can take the form of a risk register with a descriptive statement of relevant information about a potential risk. Project risk cannot be ignored, and for PFI/PPP projects, a risk register or matrix with an allocation of risk to the different parties involved is an example of a project governance tool. The absence of an up-to-date risk register and the eventual layering of risks upon risks without adequate mitigation or an effective risk management strategy were cited as one of many reasons why the multimillion pound Paddington Health Campus PFI scheme failed (NAO, 2006). Risk allocation is central in PFI/PPP projects and must be carefully managed. Akintoye *et al.* (2002) developed a comprehensive framework for risk assessment in the management of PFI/PPP projects. The tools associated with different phases of a PFI/PPP project are discussed in the following sections.

4.4   Planning and Development Phase of PPP/PFI Project

The key issues associated with the different stages in the planning and development phase of PPP/PFI projects have already been discussed in the previous chapter. This section focuses on the organisational structure and the control and monitoring mechanism to ensure good governance is achieved in the planning and development of PPP/PFI projects.

4.4.1   Organisational structure and accountability

The reporting structure of any organisation defines how power and control are cascaded throughout the organisation and are usually represented as an organograph showing reporting lines from the work streams at the bottom to the project/programme board at the head of the organograph. For PFI/PPP capital projects, simple reporting structures allow for clear accountability and decision-making. A simple structure consists of work streams coordinated and reporting to a project team and upwards to the Project Board or Steering Group which subsequently reports to the organisation's Board of Governors (DoH, 1994). The project management methodology, PRINCE2 project management manual (CCTA, 1996), gives a simple but effective structure as guidance now commonly practiced widely within, for example, the NHS (see Figure 4.1).

Complex reporting structure as shown in Figure 4.2 makes decision-making difficult and the project can easily spiral into more and more difficulties resulting in high levels of downstream risk management. Figure 4.2 is

---

[3] Quantitative risk analysis is however used in the Value for Money (VfM) appraisal of the procurement route because in PFI/PPP projects, the transfer of risk from public to the private sector is the part that normally swings the VfM in favour of the PFI route. This means that quantification of risks must be carried out.

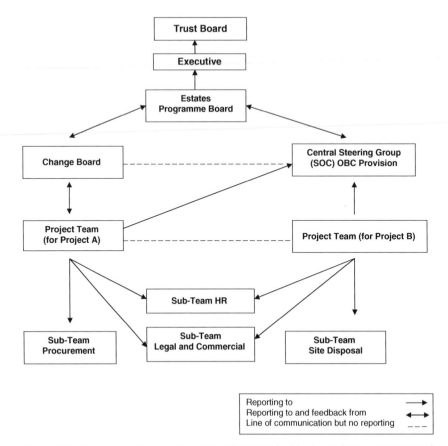

Figure 4.2 Reporting structure for a £25 million NHS build scheme (ProCure 21 project). Reproduced from Whelan (2005), with permission of Elizabeth Whelan.

an example of an ineffective project management structure from the Department of Health (DoH) with lines of responsibilities that clearly demonstrate poor control which will be a recipe for subsequent project failure. There is no sign of a project director or project manager in this reporting structure. Furthermore, the line of communication from the 'Change Board' and 'Central Steering Group' is not cascading or shown as a two-way flow of communication between the various sub-teams and project teams, suggesting poor communication exists between projects A and B and the various sub-teams.

In contrast to the above reporting structure (Figure 4.2) with complex and confusing arrows of accountability (e.g. reporting arrows from project teams to legal and commercial suggest accountability to this group). The reporting structure of a similar capital project, as outlined in Figure 4.3, is simple and effective with clear lines of accountability to the decision-making Project Board and simple overarching programming to other capital projects through the Trust's Programme Board.

Project accountability for many public organisations is defined by the reporting structure. For example, for health sector PFI/PPP projects, the line of accountability clearly rests with the Trust's Chief Executive as the SRO

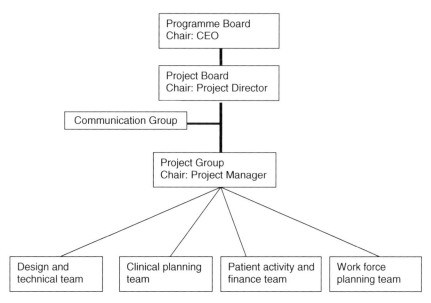

Figure 4.3   Reporting structure for a £35 million NHS specialist diagnostic centre (PPP project – procurement route undetermined).

(NHS, 1999), and for education PFI/PPP projects, it is the appropriate Board of Governors. In school PFI projects, the situation is slightly complicated as the commissioning agency, the LEA, is not the sole purchaser. There are two purchasers, the LEA and the school's governing body, but only the LEA is party to the legal contract (Edwards and Shaoul, 2003). Under these circumstances, it is important that the roles and accountability are properly defined. However, there is evidence to suggest that some control procedures, designed to ensure that only those PFI projects that have the support of the school's governing body and the other stakeholders, have not been followed (Edwards and Shaoul, 2003). In one school project, they reported that the local council/LEA prepared a business case for a school PFI project, without the active involvement of school, despite the fact that the school would ultimately be responsible for the facilities management costs or the service element of the unitary charge. In the end, the project collapsed after a few years due to inappropriate organisational structure, resulting in lack of accountability, conflicts and poor decision-making that failed to gain stakeholder consent (Edwards and Shaoul, 2003). For a project managed through poorly defined roles and responsibilities, success will be a matter of chance, and indeed it may not be possible to measure whether the project has succeeded at all.

4.4.2   Project approval processes

There are many activities associated with the planning and development phase but the key deliverables are the needs assessment, development of the strategic business case and the readiness for procurement. This initial phase of the project is also the most crucial as it determines whether the project will go through to the construction phase or be 'killed' usually at the completion of a strategic business case. A good approval process is therefore important particularly in the absence of 'front-end' resources. There is a tendency for public sector Project Boards not to invest heavily in project teams or external advisers until the project's Strategic Outline Case (SOC)[4] has been approved. A robust approval system relies on good coordination between the different teams within the organisational structure, agencies/groups outside the structure and the control and monitoring mechanism to highlight future risks before they have the opportunity to become 'red' alerts (see Section 4.4.3).

The first step, therefore, in the planning and development phase is to put the appropriate organisational structure, processes and systems as control mechanisms in place, which will ensure that required approvals and direction for the project are obtained at each stage (CCTA, 1996). The first control tool in this process is the Project Initiation Document (PID) as it contains many project governance components such as the project structure, a clear project brief, a programme, an ongoing risk register, commencement of an issue log, legal requirements and town planning papers (DoH, 2006/2007). The crucial step in terms of good project management is to have an audit trail of the documented approval process from internal board meeting minutes to external Gateway reports and other approval directives from Ministerial Departments such as, new procurement rules. These are government-directed mechanisms for public projects such as a hospital, school, and roads and are pivotal in supporting the development of a SOC. The SOC provides a case for change and an outline business case (OBC) provides justification of business.

A capital investment manual[5] (DoH, 1994) is recommended as the first point of reference for major capital projects including PPP/PFI projects in England and Wales. This manual provides practical guidance on the capital appraisal process, as well as providing a framework for establishing management arrangements to ensure that the benefits of every capital investment are identified, evaluated and realised. The CIM has recently been supplemented by additional guidance, in particular in respect of privately financed schemes such as PPP/PFI projects. A compelling business case is an essential component of project governance (Williamson, 1996). For PPP/PFI projects, the OGC provides external project review teams (described in Section 4.4.3) to that help to identify early risks to programmes and projects and ensure that public sector organisations have a robust approval process that ensure public projects are delivered on time and within allocated budget (OGC, 2004c; 2009).

The approval process is also influenced by the size of the scheme and the sector. For example, Government Guidance in 2004 on PPP/PFI projects

---

[4] The equivalent document in Scotland is the Initial Agreement (IA) produced by NHS Boards in Scotland for approval by the Scottish Government.

[5] The equivalent manual in Scotland is the Scottish Capital Investment Manual (SCIM).

stated that major capital investments with an expected capital cost of £40 million or more (an increase from the previous £25 million mark previously), will require initial approval through the appropriate Strategic Health Authority (SHA). In 2006, with the restructuring of health boundaries the number of SHA in England reduced (e.g. five SHA in London were replaced by a single SHA, NHS London) but continued in their role of being the first -line of approval for SOC and OBC. Major capital schemes have the added pressure of requiring ministerial approval following the submission of a SOC (HM Treasury, 1998).

Financial or economic appraisal tools have a vital role in the development of an OBC and are used extensively in projects to assess financial viability (Rogers, 2001). Such appraisal tools can therefore have the biggest influence on the decision to approve or not to approve a project. This is particularly important in the public sector as these decisions need to ensure that the investment approved is the best use of limited resources (DoH, 2004a, 2004b). Appraisals are routinely required and applied to support expenditure decisions involving major capital investment in public projects (London City Audit Consortium, 2005). To support organisations seeking to secure capital for future services and facilities, the Treasury's 'Green Book': 'Appraisal and Evaluation in Central Government' (1997, revised 2003) is provided for guidance on the underlying principles, methodology and use of discounted costs for public sector appraisals on capital schemes. It provides a way of expressing the total cost implications of developments over a given appraisal period (typically 60 years operation plus the construction period for the build).

The economic appraisal sections of an OBC combine both financial and non-financial appraisal of the options considered. The non-financial appraisal technique was developed to overcome the difficulty of full cost benefit analysis when so many benefits of investment are not easily quantifiable, particularly at a local level. Benefit appraisals for public projects must examine the effect of the capital project on the society as a whole. In the context of a public project, benefits can be defined as 'the economic return or advantages accruing to members of the community arising from the project' (Rogers, 2001). The final decision on 'needs assessment' will take into account not only the preferred option generated by the Generic Economic Model (GEM) but also the non-financial advantages and disadvantages of the shortlisted options. Economic analysis for a major capital project, as shown by Akintoye *et al.* (2003), should, if properly governed, demonstrate best value requirements and curtail downstream project cost escalation. The reliance on these prescribed accounting and financial methodologies to prove value for money to justify the PFI option or select the most appropriate option is an important control and monitoring mechanism (Edwards and Shaoul, 2003).

Gateway Review 1 (GW1) is the first point of project governance on the development of a sound business case for the approval of PPP/PFI projects (see Figure 4.4). This first Gateway Review focuses on the project's business justification, prior to the key decision on approval to proceed for a development proposal. Subsequent Gateway Review stages are discussed in the next Section (4.4.3).

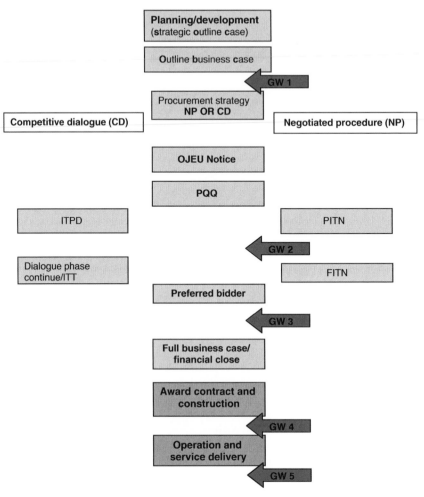

Figure 4.4 Gateway Review points for two procurement processes. *Modified from* Department of Health (DoH, 2003).

4.4.3   Project controls and gateways

Project controls are critical elements of a project that keep it to plan, on-time, and within budget and are implemented at the early planning stages right through to the benefits realisation stage. Each project should be assessed for the appropriate level of control needed; too much control being unmanageable and too little control resulting in high cost to the business when downstream changes are made to the project (Winch, 2001). The trick is to have a simple, low maintenance but robust control system that ensures alignment between project development and the organisation's broader objectives.

The government is committed to achieving excellence in public buildings and the construction of new or refurbishment projects are subject to the OGC Gateway Review Process (OGC, 2004c, 2009). This is the key control mechanism for ensuring good practice and providing confidence in project delivery. The OGC Gateway Review Process are peer reviews carried out at key decision points in a PPP/PFI programme or project, by a team of experienced people independent of the programme/project team (OGC, 2004c, 2009). The Gateway Process, as shown in Figure 4.4, provides assurance that a public build project is developing correctly and is in alignment with project objectives.

The supporting literature available from the OGC website titled, *OGC Best Practice – Gateway to Success*, is clear that the Gateway Review process is not a substitute for rigorous internal governance framework within the organisation but is best seen as being complementary to the organisation's internal governance processes. The review team examines a PPP/PFI project at five critical stages in its life cycle to provide assurance on the ability of the project to successfully proceed to the next stage by assigning Red/Amber/Green (RAG) status to the project. The coloured status reflects the level of risks the Gateway team feel have still to be addressed by the project team.

The OGC introduced new guidance to the Gateway Review process, relating to the introduction of Overall Project/Programme 'Delivery Confidence' (OGC, 2006). This is an assessment by the review team on the overall delivery confidence of the project or programme's ability to deliver the aim and objectives within the timescale, cost envelope and to the quality requirements (i.e. delivery of financial and non-financial benefits). The additional assessment will result in the removal of the RAG status for individual recommendations, and instead, each recommendation will be assigned a priority rating based on the following definition:

- **Critical (do now)**: Action must be taken with some urgency to increase the likelihood of a successful outcome.
- **Essential (do by)**: Action must be taken in the near future to increase the likelihood of a successful outcome.
- **Recommended**: The programme/project should benefit from the uptake of this recommendation.

Following the approval at Gateway Review 1 (GW1), a procurement strategy is developed; the project is advertised to generate interest from potential PPP/PFI contractors. The Pre-Qualification Questionnaire (PQQ) establishes the bidders' ability, track record, financial soundness and technical capacity for the proposed construction and facilities management (OGC, 2003). Rigorous assessment and scoring/benchmarking of these qualities at PQQ stage is crucial. Initially, the process was to a large extent subjective and very much dependent on the public sector client's professional advisers (OGC, 2003), and hence, the importance of appointing competent advisers. A PFI guidance note, Technical Note 3 (Treasury, 1998), was developed by the Treasury Taskforce (TTF) as a direct response to the recommendation by Sir Michael Bates during his review of the PFI process (HM Treasury, 1997). The TTF, itself formed in direct response to the Bates review, was tasked to develop a means of improving the quality of advisers by, first, checking credentials and testing their knowledge, commitment and depth of resource and, second, clarifying the role of advisers so as to ensure consistency throughout PFI/PP projects. Standardised PQQ/templates have also been introduced particularly in mature PFI/PPP sectors such as health and are available on various OGC websites (OGC, 2009). The standard PQQ, the rigorous assessment process and various PFI guidance notes by the TTF are all examples of control and monitoring mechanisms.

Since 31 January 2006, all complex contracts such as PFI/PPP projects are subject to a new European Union public sector procurement directive (2004/18/EC), called the 'competitive dialogue'. It is important to note that there are three potential areas of infringement between the old negotiated procedure and the new rules for competitive dialogue (OGC, 2006):

- Sanction of the use of the negotiated procedure
- Selection of a reserve bidder
- Awarding preferred bidder status to a candidate subject to a number of conditions or reserved matters

The selection process for competitive dialogue, following the expression of interest, is carried out in accordance with the European Union public sector procurement directive (Articles 44–52). These articles also cover the selection process for restricted and negotiated procedures.

For the competitive dialogue procedure, the next step is an Invitation to Participate in Dialogue (ITPD), and it is the starting point to competitive dialogue with all candidates. The number invited can be reduced to three, provided this is sufficient to ensure effective competition. For negotiated procedure, the objective of the preliminary invitation to negotiate (PITN) is to distil a list of six potential bidders 'who can undertake a project like this' to three 'who can demonstrate that they can do this project' (OGC, 2003). Tenders from no more than three bidders are selected by the contracting authority for further negotiation. The market forces generated through competition amongst bidders at these stages 'provide a measure of control' and the 'best assurance for value for money' (Edwards and Shaoul, 2003). The second point of project governance through Gateway 2 (GW2) is exercised here at the early stages of competitive procurement and involves the review team approving the public sector client's proposals. This review has a focus on the following:

- General management of the procurement process (including having the right skill-mix in the integrated design team).
- For competitive negotiation, the sufficiency of the requirement for design information from bidders to enable and evaluate their proposals in accordance with design and environmental tools such as the 'Achieving Excellence in Design Evaluation Toolkit' (AEDET) and 'BRE Environmental Assessment Method' (BREEAM), respectively.
- For competitive dialogue, design information and requirement should be included in the contract notice or in a 'descriptive document'[6] before engaging in procurement.

---

[6] The term 'descriptive document' is used in competitive dialogue (CD) in contrast to the use of 'specification' to cover the broader approach in CD of setting out needs and requirements for which different solutions will be proposed.

Once the awarding authority is satisfied that the dialogue phase can provide a solution or a number of solutions satisfying the project requirements, bidders in the dialogue are requested to submit their final tender offer during the Invitation to Tender (ITT) stage for evaluation. Once the final offers have been tendered, limits are placed in terms of discussions allowed and therefore, the final offers must contain all elements required to deliver the contract. The subsequent stage from PQQ for the negotiated procedure is known as the Final Invitation to Negotiate (FITN) stage and is primarily to request more detailed information from up to three shortlisted bidders on planning and architecture, engineering services and other information such as service level cost to determine affordability. For the public sector clients, the task is to reduce three bidders to one. In some cases, there is a modification of the procurement process, as the PITN and FITN are combined as a single ITN. This will normally be the case where there is a very large project with few 'players' in the market. Under these circumstances, the PQQ process may be strengthened to arrive directly at the shortlist.

Control mechanisms for guiding actions of decision-makers (i.e. members of the Project Board) during the dialogue phase or final negotiation stage are usually provided via a participatory process resulting in a peer-reviewed evaluation report that will set out the performance of each bidder against the selection criteria. At this stage, the project manager will put forward a recommendation. Under the negotiated procedure, this will involve two bidders, having had advice from external consultants and the Private Finance Unit (PFU)[7].

The preferred bidder is selected via the competitive dialogue procedure by evaluation against predetermined weighted award criteria, in line with Best Practice and European Commissioning case law. For the negotiated procedure, following the submission of fully developed bids, the preferred bidder is selected from the remaining two bidders. This may be achieved by giving no more than two bidders an opportunity to revise their proposals along the lines agreed with the public sector client – and their prices.

Choosing the preferred bidder is a decision closely worked up with the representing government departments, the Head of PFU (England) or PFCU (Scotland), to ensure the selected bidder represents value for money and affordability. This guidance precedes any recommended endorsement from the Project Board.

The third Gateway Review (GW3) is carried out once a preferred bidder has been selected and forms part of the contracting process between the public sector client and preferred supplier or private sector service provider. This review investigates the full business case (FBC) and the governance arrangements for the investment decision. This is a final check to confirm the project is still required; affordable and achievable; implementation plans are robust and investment decision is appropriate. To avoid abuse/wrong decisions and to derive the best benefit as a project control step, this review takes place before a work order is placed with a supplier, funding and resources committed.

---

[7] The equivalent in Scotland is the Private Finance and Capital Unit (PFCU).

4.4.4   Post-project evaluation

The importance and value of post-project evaluation are exercised during the development of the full business case, with the preferred bidder, after contract is awarded for construction. A key requirement for approval in the health sector PFI/PPP projects is that the Trust's Chief Executive, the relevant governing SHA and the DoH are all satisfied that adequate provision has been made at the outset to undertake post-project evaluation. Gateway Review 4 (GW4) is pivotal to the evaluation process as it ensures that the organisation is ready to make the transition from the construction phase and that ownership and governance are in place for the operational phase. The review focuses on the readiness of the organisation to go live with the necessary business changes, the management of the operational services and an understanding that the benefits realisation plans are likely to be realised.

It is important that lessons learned during post-project evaluation are appropriately documented for knowledge transfer to address problems in future PFI/PPP projects. The success and failures of PPP/PFI projects have been widely reported (Birnie, 1999; Akintoye et al., 2003) often by political groups with their own particular opinion who wish to use it to fit their own agenda. The result is a series of mixed messages and often bad media reports inevitably follow some of the more highly political and publicised schemes. For example, the 'Times' newspaper compared the 'Everlina Children's Hospital' (a non-PFI project) with the Royal London NHS Trust development in East London. It described the former as a 'public delight' and the latter (quoting Commission for Architecture and the Built Environment (CABE)) as 'cramped', 'confusing' and seriously 'flawed' (CABE, 2005). Nevertheless, the government is confident that it is possible to achieve excellence in design through the PFI process and has promoted several initiatives, for example encouraging the establishment of programme centres and guides for design quality (OGC, 2004a). However, it is to be noted that the government's own design watchdog (i.e. CABE) has argued that the government's expectation of design innovation from private sector providers has not been forthcoming (CABE, 2002).

Post-project evaluation provides an opportunity to confirm the key benefits of PFI/PPP projects and to identify areas for improvement. Such benefits include opportunity of better value for money; proper focus on whole life costing; fully integrating upfront design and construction costs (Birnie, 1999; Akintoye et al., 2003; OGC, 2004b). In addition, PFI/PPP ensures an integrated supply chain is in place from the earliest stages of the design process and provides wider opportunity and incentive for innovative solutions as to how service requirements can be delivered (OGC, 2004b).

It is now 12 years since the first PFI projects have been built and are in service. To date, improvements in the process have concentrated on design, contractual issues and payment mechanisms, and the more the projects that are delivered, the better served the process is. The HM Treasury paper titled 'PFI: Strengthening Long-Term Partnerships' (HM Treasury, 2006) identifies a number of ways in which the government could improve the PFI/PPP process. One suggestion involves developing a secondment model within the public sector so that public servants with experience of complex procurements can be retained and deployed on projects across the public sector. Another suggestion is to take steps in the forthcoming comprehensive spending review to ensure that PFU are appropriately resourced to manage their PFI programmes and a further suggestion is to develop individuals and teams procurement skills through formal qualification and training.

The policy for successful delivery was developed to improve public services effectiveness at delivering programmes and projects. The complete policy package of successful delivery toolkit describes best practice principles that have been proven to work, leading to improved performance and outcomes (OGC, 2003). The aim is to bring together in a single point of reference both policy implementation and best practice for managing projects at different stages of PPP/PFI from the planning and development phase to the operational phase (OGC, 2002).

4.5   Construction Phase of PPP/PFI Projects

A significant part of the design would have been completed and reviewed during the planning and development phase of PPP/PFI projects but it is important to understand the key control and monitoring mechanisms to achieve design quality and improve construction performance. This is crucial as design quality influences the construction process.

4.5.1   Design controls for construction

Design development underpins all PPP/PFI projects, including those perceived as simple. The PFI/PPP of the 'design development' process is critical within the chosen procurement process because it is at this point of the design stage that most can be done to optimise the 'value' of the finished facility to its end users. Hence, in achieving optimum design quality, public sector client organisations undertaking a PFI/PPP scheme must observe the design development protocol (DoH, 2004c). This is a prescriptive process that runs concurrently with the PFI/PPP planning and development/procurement process, as shown in Figure 4.5. To support the planning and development phase, there is no shortage of governance tools and techniques such as project reviews (OGC, 2004c, 2009) and 'Achieving Excellence in Design Evaluation Toolkit' (AEDET; DoH, 2006). Figure 4.5 is an outline of the design development process alongside the key stages in the planning and development phase. This diagram indicates where two governance tools, AEDET and the Gateway Reviews, have an important governance role, prior to a project proceeding to the next design stage.

The design development protocol (2004c) is shown in Figure 4.5 running concurrently with the negotiated procedure procurement process but it can be adapted to run in parallel with the new procurement process, the competitive dialogue procedure, by simply replacing PITN in the above diagram (Figure 4.5) with ITPD and replacing FITN with the dialogue phase.

Procurement process   **Design process**

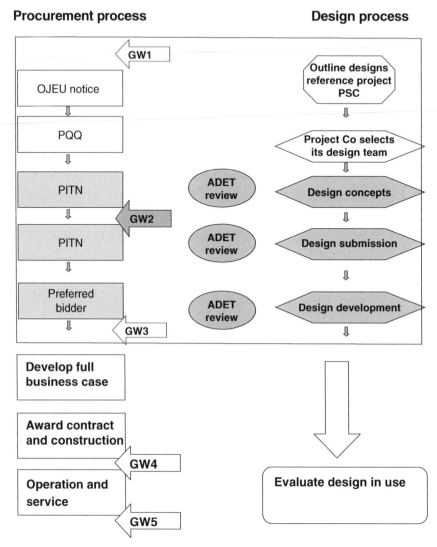

Figure 4.5 Design control process running parallel with procurement stages. *Source*: Adapted from DoH (2004c).

4.5.2   Project management

An appropriate project management structure is crucial to speed up decision-making during the implementation of PFI/PPP projects. The rigorous pre-qualification and the bid evaluation process at the planning and development phase acts as a powerful control mechanism for the commencement of construction as it ensures that the project management structure, quality of leadership and the team selected is the most appropriate for aligning and leading the construction programme. Project managers have been identified as the single most critical factor affecting successful project delivery

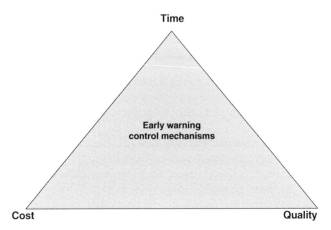

Figure 4.6   Control mechanisms (EW) at the centre of a project management triangle. *Source*: Modified from Health facilities Scotland, NEC Training, December 2008.

(Hartman, 2000; Bandow, 2001; Powl and Skitmore, 2005). This remains the case for PFI/PPP projects which are particularly complex and must be tested during the evaluation process. Professionals involved in the delivery of PFI/PPP projects will be aware of projects which have failed, because of ineffective project management structure and poor leadership. For project managers to be effective in the delivery of build projects, they need to be nurtured and encouraged (Pinto and Slevin, 1989); be generalists rather than specialists (Pinto and Kharbanda, 1995). They also need to work within a system that encourages creativity and innovation (Ramo, 2002), do 'the right thing at the right moment' (Ramo, 2002) and avoid ineffective traditional ways and bad practices (McKenna, 1998). The appointment of a competent project manager with appropriate qualities is therefore a critical factor to successful project delivery of PFI/PPP projects.

Cartlidge (2004) identified other reasons for project management failure such as projects costing more than the originally planned budget, late delivery or delivery of an inferior product. Traditionally, these constraints of quality, time and cost are often referred to as the Project Management Triangle (PMT, Figure 4.6). Each side represents a constraint that is often a competing constraint. For example, increased project quality scope usually means increased time and increased cost. A tight time constraint could mean increased costs and reduced quality or project scope, and a tight budget could mean increased time and reduced quality scope (Westerveld, 2003). Early warning (EW) control mechanisms and appropriate risk allocation are powerful control mechanisms for cost, time and quality in PFI/PPP projects. The EW mechanism is an essential project management tool for alerting the construction team at the earliest opportunity that an issue if not corrected urgently will have an impact on overall project delivery. EW control mechanisms are therefore at the centre of the PMT as shown in Figure 4.6.

Under PFI/PPP projects, the risk of time overrun, cost overrun and quality are transferred to the private sector contractor. The implication is that

the PFI contractor has to adopt effective tools and techniques for controlling the project. The PFI/PPP contractor does not receive any payment until
the project is completed and becomes fully operational. There are also payment deductions for service delivery failures arising from the quality of the
asset due to poor project management during the construction phase. The
project management structure is absolutely fundamental, so too is the need
to have available project management tools and techniques that enable the
project team (not just the project manager) to organise their work and meet
these often conflicting constraints (Clarke, 1999) and to avoid cost and time
overruns.

To implement projects successfully during the construction phase, it is also
important to understand the impact of critical success factors and project success factors. Project success factors (PSF) are those factors that are required
and necessary for successful project execution such as improved team communication, focus and energy. Critical success factors on the other hand are
those factors that are critical to the overall project delivery (Belassi and Tukel,
1996; Naoum et al., 2004). A number of studies in the project management
literature concentrate on the critical factors that affect project success or failure (Belassi and Tukel, 1996; Walker and Naoum, 1997; Chan et al., 2004;
Hughes et al., 2004). Whilst many of these studies generate lists of critical
success factors (CSF), each list varies in scope and purpose. Most of the early
studies focused on the reasons for project failure rather than project success
(Avots, 1969; Morgan and Soden, 1979; Hall, 1980). In these studies, it was
assumed, rightly or wrongly, that if a project's completion time exceeded
its due date, or expenses overran the budget, or outcomes did not satisfy a
client's predetermined performance criteria, then the project was assumed to
be a failure. However, determining whether a project is a success or failure is
more complex in PPP/PFI projects where the risks of time and cost overruns
are transferred to the private sector PFI/PPP contractor.

The PSF are not significantly different for the construction phase of PPP/PFI
projects as the processes are similar. Tukel and Rom (1995) argue that the
most critical factor for the successful completion of projects is top management support and this should be clearly reflected in the project's organisational structure. The support is usually strongest if there is a project
champion from the executive team who is then able to support project managers understand and achieve the project objectives. This is certainly true,
for example in major hospital development PFI/PPP projects where a clinical
champion is seen to be pivotal to achieving clinical buy-in to the proposed
change (Greenway et al., 2007). Pinto and Slevin (1989) also found that
environmental factors such as political and economic factors affect projects
during the planning stage of a project cycle; the Paddington Health Campus
and the Royal London and Bartholomew Hospital PPP/PFI projects are two
good examples of political influence affecting projects during planning stages
(NAO, 2005b, 2006).

More recent work by Li et al. (2005) classified 18 CSF for the delivery
of construction PFI projects. This research study grouped success factors
into five basic elements for successful project delivery, determined their relative importance to one another, and subsequently, ranked them in order of

importance. From a list of 18, good governance was ranked as the ninth most important CSF for PPP/PFI construction projects, strengthening the argument for the importance of governance in relation to project delivery. The study concluded that five factor groupings represented the basic elements of CSF for PFI project development, and should always be considered by public sector sponsors in forming and shaping their PFI policy development, and by private sector PFI/PPP contractors in managing their projects. The five factor groupings identified are effective procurement, project implementability, government guarantee, favourable economic conditions and available financial markets. In terms of governance, its importance for PFI was highlighted in terms of developing sound economic policy, administering and facilitating project delivery.

4.5.3   Project performance

Key performance indicators (KPI) are measures that quantify objectives and enable the measurement of operational performance (Belassi and Tukel, 1996). For PFI/PPP projects, KPI measure service delivery and the unitary cost is reduced where the operator does not perform. There are a number of initiatives to improve performance of construction projects including KPI, benchmarking to compare performance between varieties of strategically important performance criteria (Carr and Winch, 1998). Work by Robinson *et al.* (2005) demonstrated that construction organisations are keen to benchmark their knowledge management activities in an effort to improve project performance.

Performance indicators can be used to measure the effectiveness and efficiency achieved relative to a desired function/output and these can be expressed in terms of time, cost, quality or a combination (Kemps, 1993). Closely related to performance indicators are the concepts of 'benchmarking' and 'best practice'. The government in its quest to achieve excellence in construction of public buildings (OGC, 2004c) and construction of new or refurbished hospitals now subjects all public capital projects to an assessment of project performance. A study by Belassi and Tukel (1996) describes the impact of PSF (e.g. common methodologies, use of monitoring tools and experienced project managers) on project performance. Interestingly, many of the success factors described by Belassi and Tukel (1996) are also elements of project governance. A programme management survey by KPMG in 2002 (OPSR, 2003) examined the impact of a programme office (a department in the organisation responsible for managing a programme of projects) and found there was a strong correlation between a Programme Office (PO), effectiveness/maturity and project success.

Best and De Valance (1999) argues there are three critical areas of assessment for any facilities: physical performance, functional performance and financial performance. For example, in public buildings such as hospitals, physical and functional performances are assessed by the government-directed tool AEDET which is now an integral part of PFI/PPP projects as discussed in Section 4.5.1. This is a design quality specialist tool produced for evaluating the design of buildings and has three main categories: functionality, impact and build standard. These are further subdivided into ten categories: use, access, space, character and innovation, public satisfaction, internal environment, urban and social integration, performance, engineering and construction. A further tool, A Staff and Patient Environment Calibration Tool (ASPECT), has been developed by the DoH in England to address eight key health care design issues: privacy, company and dignity, views, nature and outdoors, comfort and control, legibility of place, interior appearance, facilities and staff. ASPECT can be used as a stand-alone tool, or it can be used to support AEDET evolution to provide a more comprehensive evaluation of the design of health care environments (OGC, AEDET and ASPECT evidence layer, December 2007).

In PPP/PFI projects, functional performances are crucial from the perspective of the public sector client and private sector operator. Failures in how the facilities function could affect service delivery resulting in a deduction in unitary payments from the public sector client. Financial performance, in terms of whole life cost and operational cost, could also affect the profitability of PFI/PPP contractors. Public assets have not been properly maintained in the past, due to tight financial constraints often resulting in cutbacks on maintenance spending (Ball and King, 2006). PFI projects encourage the private contractor to consider costs over the whole life, and the lower costing from the PFI/PPP consortium is due to the strong incentives to 'reduce costs but not to jeopardize quality' as there are penalties associated with this. In schools, PFI/PPP projects under the Building Schools for Future (BSF) programme, timescales and costs are controlled by local authorities and Partnership for Schools (PfS). The local education partnership or the PPP contractor procures the delivery of the project through a benchmark and market-tested supply chain. Once construction is completed, the facility is then tested against planned objectives through a mechanism known as 'post-occupancy evaluation' to assess whether benefits at the planning and development phase are realised.

The Treasury has, for many years, provided guidance to public sector bodies on how proposals should be appraised in terms of financial performance, before significant funds are committed. The new revised edition of the Treasury Green Book (1997; revised 2003) aims to encourage a more thorough, long-term and analytically robust approach to appraisal and evaluation. For example, health sector projects are benchmarked against updated NHS departmental cost allowances during economic appraisal stage of the business case at Gateway 3 (see Figure 4.4).

Mott MacDonald (2002) in his review of large public sector projects in the United Kingdom concluded that the poor performance was rooted to the planning and design team's optimism with respect to risk when appraising the projects. Among the recommendations of the Mott MacDonald report (2002) is the need for effective project controls such as methodical archiving of key documents and an open approach to sharing the successes and failures of major projects through internal and external seminars and post-project reviews.

4.6   Operation and Service Delivery Phase of PPP/PFI Project

The operational and service delivery phase is where the value-for-money argument is tested according to PFI theory. It is therefore crucial to have appropriate monitoring and control mechanisms to ensure that value for money is not undermined in the delivery of services. Under the Capital Financial Regulations (1997), a contract in PFI should be paid when services are provided to meet the client's requirements. Payments in PFI/PPP projects are based on performance in service delivery which is determined by the output specification of the public sector client.

The level of services proposed or agreed by the private sector operator is finalised in the full business case and is subject to the final point of project governance through the Gateway Review Team. Gateway Review 5 (GW5, Figure 4.4) is repeated throughout the life of the contract ensuring benefits are achieved and performance and VFM are maintained. This review confirms that the desired benefits of the project are being achieved, and the business case changes are operating smoothly. The organisational structure, control and monitoring mechanisms used at this stage are discussed in Section 4.6.1.

4.6.1   Control and monitoring of service delivery

There are several elements required for control and monitoring service delivery. Firstly, the output specification in a PFI/PPP project defines the client requirements and project scope, and is reflected in the business case during the planning and development phase. The output specification sets out the requirements of a project in terms of accommodation standards and services. Services cover a wide range of hard and usually soft facilities management services such as maintenance, landscaping, infection control, cleaning, catering, security, and so on. The performance standard should define the threshold for pass or fail in order to assess the level of service compliance and the rectification period allowed if service fails, taking into account certain relief events, for example lack of access. These mechanisms ensure a proper control and monitoring of service delivery.

Standard documents or templates for a wide range of facilities management services are now available particularly for mature PFI/PPP sectors such as health. A well-drafted output specification is therefore fundamental to the successful delivery of long-term service delivery. Appendix A is a good example of an output-based specification (OBS) for a generic ward and was one of many control documents for a multi-million pounds PFI project. Examples of OBS for facilities management service standards are shown in Appendix B.

Secondly, it is important to determine how the output specification for accommodation and services are related to the performance scoring and payment mechanism. The unitary charge or payment mechanism in PFI/PPP projects combines all the different elements of accommodation and service as the basis of repayment to the PFI contractor. Method of measurement should also be agreed between service provider and public sector client. See

---

Box 4.1: Example of KPI

The performance measurement scores are all calculated out of 100%.

The performance standards (KPI) have been allocated a direct measurement principle. Measurement principles are ones where it is possible to measure performance standard over a measurement month accurately.

Where performance target is less than 100%, then performance will be measured by reference to this target. For example, target is 95% and measured performance is 96%, then the standard score will be 100%. If the measured performance is 90%, then the standard score is calculated as $(90/95) \times 100 = 94.74\%$.

For example, Performance Standard 1.1.2 lists 30 events via the 'Help Desk' for which 3 were not responded to within the required response time. Performance failure would be 10% $[(3/30) \times 100]$.

Therefore, the standard score for the month is 90%.

---

example of how performance measurement scores are calculated in box 4.1. The payment mechanism depends upon level of services and the criticality of the accommodation facility which varies between different facilities, for example classroom, operating theatre, computer room and laboratory. This is captured in the payment mechanism as it reflects the size of payment deduction should a particular facility not be available. The payment formula acts as a control and monitoring mechanism to regulate the behaviour of the private sector operator as it puts into financial effect the allocation of performance and operational risks during the delivery of services according to the output specification. There are various payment models for PFI/PPP projects (discussed in Chapter 3) which allows for an annual adjustment for inflation through indexation and also for periodic adjustments for the service component of the unitary charge through benchmarking and market testing (Boussabaine, 2007). Indexation, benchmarking and market testing are examples of further control and monitoring mechanisms to ensure that value for money is achieved throughout the operational stage. In many PFI/PPP projects, the performance of FM services is measured according to a percentage scale. If the standard requirements are met, the amount paid to the contractor will be reduced according to the level of service failures.

Value-for-money arguments at the planning and design development phase depend on good governance such as measurement and audit mechanisms during the operational phase. It is the level of compliance with the output specification, both in terms of the measurement scores for availability and the standard of service which determines the payment due from the public sector client. Availability payment relates to the provision of assets such as buildings/equipment whilst service payments are for facilities management (FM) services. Performance measurement works on several levels. Performance is scored against services or functional areas, and the payment mechanism calculates deductions based on the level of service failure or unavailability of functional areas. Each functional area is made up of functional units. For example, a ward is made up of a number of bedrooms, staff room, storeroom, cleaner's room, and so on. Robinson and Scott (2009) cited an example of a larger PFI hospital which has 49 functional areas and 1200 functional units. Non-performance in functional units or areas can lead to lower scores and payment deduction. Each service is also subdivided into 'scopes of service' which is then divided into 'aspects of service' and each aspect has one or more 'performance standards' to be achieved. The performance standards are measured in one of three ways based on KPI (see Appendix B2 for examples), tariff or auditing systems as shown in Boxes 4.1–4.3.

Box 4.2: Example of Tariff Measurement Scores

The performance tariff measurement scores are all calculated out of 100%.

Tariff measurement scores are those where a small failure to comply will have a serious impact on service provision. For example, Performance Standard 6.2.3 requires fuel supplies to be available at all times (other than under particular circumstances). The agreed tariff for this standard is as follows:

1. Unavailable 0–1 hour – no deduction from 100%.
2. Unavailable for more than 1 hour – 20% deduction from 100% for each unrelated occurrence.
3. A further deduction from 100% based upon total time lost in the month.
   Incident 1 – time lost 0.5 hour gives 0% deduction from 100%
   Incident 2 – time lost 6 hours gives 20% deduction from 100%
   Incident 3 – time lost 12 hours gives 20% deduction from 100%
4. Additional deduction related to time lost (assume 30-day month) $[(12 + 6 + 0.5)/(24 \times 30] \times 100\% = 2.6\%$.

Therefore, standard score using tariff measurement principle $= 57.40\%$ $[100\% - (20\% + 20\% + 2.6\%)]$.

Box 4.3: Example of Performance Audit Scoring System

The performance audit scores are all calculated out of 100%.

Auditing will be carried out in accordance with Section (d) in each part of Schedule 8. The precise details of frequency and monitoring checklists will be set down in the site-specific service quality plans prepared by the service provider. Performance will be assessed by reference to three basic types of monitoring:

Supervisory: Carried out routinely by the direct supervisor as part of the supervisory role.

Managerial: Carried out by the site-based managers of the service provider as part of the overall management role.

Internal audit: Carried out by in-house QA auditors of the service providers and the service quality manager.

The audit score is calculated as follows:
   $AS = (AI - FI)/AI \times 100$
which will give the audit score in percentage, where
AS = Audit score
AI = Total number of audited items/or total audited area
FI = Total items or amount of area which failed to achieve performance standard (after rectification period allowed)

When the performance audit failure is in the range 0–15%, the standard score will be 100% less a percentage that is double the performance failure; for example if the performance failure was 2%, the standard score would be $100\% - (2\% \times 2) = 96\%$.

When the performance audit failure is in excess of 15%, the standard score will be (i) 100% less and (ii) 30% a percentage that is treble the difference between the percentage of failure and 15%. For example, if the performance failure was 20%, the standard score would be $100\% - 30\% = 70\% - (20\% - 15\%) \times 3 = 55\%$.

Each PFI project contains performance targets weighted against the disruption that a service lapse would cause. Performance measurement and audit system therefore provide an effective control mechanism to ensure value for money so that the PFI contractor delivers according to the output specification.

Thirdly, a monitoring mechanism is needed to provide incentives and sanctions for the service provider to deliver the level of services stipulated in the output-based specification. There are a range of methods used to measure performance in PFI/PPP projects. For example, in prisons, a PFI/PPP project's performance is monitored by an on-site controller, often recruited from the operational part of the prison service and who reports to the Commissioner for Correctional Services (NAO, 2003). In the health sector, project performance is measured against the output-based specification and uses patient satisfaction surveys, patient flow audits and spot checks. The critical performance review will take place during the OGC Gateway 5, Operations review and benefits realisation.

When the service provider is using a self-monitoring method, this is usually subjected to a regular audit by the public sector body. In some cases, the service provider monitors the performance of its own subcontractors which is then subject to monthly audits and random checks by both the SPV/SPC and the public sector body. In a recent survey of local authority operational PFI, 85% of respondents stated that monitoring reports were being prepared by the service provider and assessed by the local 'authority for each payment period' (4Ps, 2005).

Apart from payment deductions, there are also other effective sanctions that are triggered by different events as shown in Figure 4.7.

Principal–agent theory asserts that 'increased performance incentives *ceteris paribus* raise the agent's productivity' (Kunz and Pfaff, 2002). Robinson and Scott (2009) cited some cases where performance systems are complex and there is a lack of understanding amongst public sector staff. The performance measurement, monitoring system and payment mechanism should therefore be designed not to be overly complex but to provide a strong incentive and controls for the PFI contractor (agent) to put more efforts to deal with service lapses.

4.6.2   Organisational structure

Partnerships UK (2006) found that there is some evidence to suggest that those projects with dedicated contract management report a higher level of improved performance. Partnerships UK (2006) also reported that over 70% of projects surveyed are managed by formal contract management teams; although just over half the respondents devote two or less people to the day-to-day contract management. In some PFI/PPP projects, there are committees that meet monthly to discuss performance and deal with issues raised. Such a committee may consist of users' representatives, the SPV/SPC, the FM service provider and the public sector client. There may also be quarterly liaison meetings attended by higher level representatives from the SPV/SPC, service

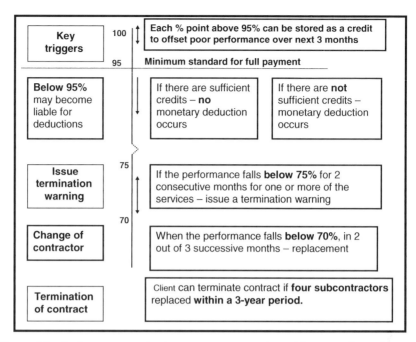

Figure 4.7 Performance monitoring system. *Source*: Adapted from NAO (2005a).

provider, client and other stakeholders focusing on more long-term service delivery issues. Recent study by Robinson and Scott (2009) suggests that the public sector has not fully assessed the resource implications of performance monitoring and consequently have not set aside sufficient resources. The study also demonstrated the importance that all parties placed on partnership between the public and private sector in discharging their responsibilities. Staff changes were raised as an issue that can interrupt efforts to build effective working relationships (Scott and Robinson, 2008). The need for an independent organisation or third party paid jointly by the SPV/SPC, service provider and public sector client to audit and certify performance is increasingly recognised as important.

4.7   Concluding Remarks

The government has repeatedly endorsed PFI/PPP as the favoured method for meeting the government's building and infrastructure programme particularly in sectors such as health and education. The key issues facing public sector clients is making an economically sound business case for PFI projects and streamlining the PFI procurement process to rapidly deliver value-for-money public projects with transparent governance and accountability arrangements.

This means having in place effective reporting structure, project control mechanisms and competent project management processes. To aid this process, there is no shortage of government directives or policies on governance and management of capital projects in the public domain. These directives come in the form of 'best practice' guidelines and government-directed toolkits. However, there are issues of implementation of these governance tools to ensure public projects are successfully delivered. Equally, despite the wealth of information available, many complex and large-sized projects are 'one-off', and therefore, it is essential that components of governance such as project approval, controls, risk management and accountability be carefully considered in relation to project delivery. These issues will be addressed in the case studies found in subsequent chapters.

References

4Ps (2005) *4Ps Review of Operational PFI and PPP Projects.* Available online at http://www.4ps.gov.uk (Accessed 14 March 2008).

Akintoye, A., Beck, M., Hardcastle, C., Chinyio, E., and Asenova, D. (2002) Framework for risk assessment and management of private finance initiative projects. *Final Report EPSRC/DTI.* Glasgow Caledonian University.

Akintoye, A., Hardcastle, C., Beck, M., Chinyio, E., and Asenova, D. (2003) Achieving best value in private finance initiative project procurement. *Construction Management and Economics* 21, 461–470.

Avots, I. (1969) Why does project management fail? *California Management Review* 12, 77–82. [Cited in: Belassi, W., and Tukel, O.I. (1996) A new framework for determining critical success/failure factors in projects. *International Journal of Project Management* 14(3), 141–151.]

Badshah, A. (1998) *Good Governance for Environmental Sustainability, Public Private Partnerships for the Urban Environment Programme (PPPUE).* United Nations Development Program (UNDP), New York.

Ball, R., and King, D. (2006) The private finance initiative in local government. *Economic Affairs* 26(1), 36–40.

Bandow, D. (2001) Time to create sound teamwork. *Journal for Quality and Participation* 24(2), 41–47.

Belassi, W., and Tukel, O.I. (1996) A new framework for determining critical success/failure factors in projects. *International Journal of Project Management* 14(3), 141–151.

Best, R., and De Valance, G. (1999) *Building in Value: Pre-Design Issues.* Arnold, a member of the Hodder Headline Group. Available online at http://www.arnoldpublishers.com.

Birnie, J. (1999) Private finance initiative (PFI) – UK construction industry response. *Journal of Construction Procurement* 5, 5–14.

Boussabaine, A. (2007) *Cost Planning of PFI and PPP Building Projects.* Taylor and Francis, Oxford.

CABE (2002) *Press Release.* Available online at http://www.cabe.org.uk (Accessed 18 September 2006).

CABE (2005) *News, Final Views on the Royal London Hospitals Proposals.* Available online at http://www.cabe.org.uk (Accessed 18 September 2006).

Carr, B., and Winch, W. (1998) *Construction Benchmarking: An International Perspective.* Bartlett Research Paper 3, University College London, London (UK), Funded by Engineering and Physical Research Council.

Cartlidge, D. (2004) *Procurement of Built Assets.* Elsevier Butterworth-Heinemann, Oxford.

CCTA (1996) *Managing Successful Projects with PRINCE2*. HMSO, United Kingdom Official Publications Database, London.

Chan, A.P., Scott, D., and Chan, A.P.L. (2004) Factors affecting the success of a construction project. *Journal of Construction Engineering and Management* 130(1), 153–155.

Civil Service Governance (2008) *Governance Web Page*. Available online at www.civilservice.gov.uk.

Clarke, A. (1999) A practical use of key success factors to improve the effectiveness of project management. *International Journal of Project Management* 17(3), 139–145.

Corkery, J. (1999) *Governance: Concepts and Applications*. International Institute of Administrative Sciences, Brussels.

Department of Health (DoH) (1994) *Capital Investment Manual*. HMSO, United Kingdom Official Publications Database, London.

Department of Health (DoH) (revised 2004a) *Guide to Using the Excel OBC Generic Economic Model*. HMSO, United Kingdom Official Publications Database, London.

Department of Health (DoH) (revised 2004b) *Principles of Generic Economic Model for Outline Business Case Option Appraisal*. HMSO, United Kingdom Official Publications Database, London.

Department of Health (DoH) (2004c) *The Design Development Protocol for PFI Schemes – Revision 1*. HMSO, United Kingdom Official Publications Database, London.

Department of Health (DoH) (2006) *Achieving Excellence Design Evaluation Toolkit (AEDET Evolution)*. HMSO, United Kingdom Official Publications Database, London.

Department of Health (DoH) (2006/2007) *Good Practice Guidance: Public Private Partnership in the National Health Service: The Private Finance Initiative*. Available online at http://www.dh.gov.uk/ProcurementAndProposals/PublicPrivatePartnership/PrivateFinanceInitiative/PFIGuidance/fs/en.

Edwards, P., and Shaoul, J. (2003) Controlling the PFI process in schools: a case study of the Pimlico Project. *Policy and Politics* 31(3), 371–385.

Greenway, B., Salter, B., and Hart, S. (2007) How policy networks can damage democratic health: a case study in the government of governance. *Public Administration* 85(3), 717–738.

Hall, P. (1980) *Great Planning Disasters*. Weidenfeld and Nicolson, London. [Cited in: Belassi, W., and Tukel, O.I. (1996) A new framework for determining critical success/failure factors in projects. *International Journal of Project Management* 14(3), 141–151.]

Hartman, F.T. (2000) The role of trust in project management. *PMI Research Conference*, Alberta, Canada.

HM Treasury (1997) *Review of the PFI by Sir Malcolm Bates, Summary and Conclusion*. HMSO, United Kingdom Official Publications Database, London.

HM Treasury (1997, revised 2003) *Appraisal and Evaluation in Central Government: The Green Book*. HM Treasury, London.

HM Treasury (1998) *A Step-By-Step Guide to the PFI Procurement Process* (revised edition). HM Treasury, London.

HM Treasury (2006) *PFI: Strengthening Long-Term Partnerships*. HM Treasury, London.

Hughes, S.W., Tippett, D.D., and Thomas, W.K. (2004) Measuring project success in the construction industry. *Engineering Management Journal* 16(3), 31–37.

Kemps, R. (1993) *Fundamentals of Project Performance Measurement*. San Diego Publishing Company, San Diego.

Kunz, A.H., and Pfaff, D. (2002) Agency theory, performance evaluation and the hypothetical construct of intrinsic motivation. *Accounting, Organisations and Society* 27, 275–295.

Latham, M. (1995) *Getting the Act Together*. Building Magazine, 1 December, London.

Li, B., Akintoye, A., Edwards, P.J., and Hardcastle, C. (2005) Critical success factors for PPP/PFI projects in the UK construction industry. *Construction Management and Economics* 23, 459–471.

London City Audit Consortium (2005) *Guide to the Corporate Governance and Internal Audit of Major NHS PFI Schemes*. HMSO, United Kingdom Official Publications Database, London.

MacDonald, M. (2002) *Review of Large Public Procurement in the UK*. HM Treasury, London.

McKenna, P.J. (1998) Play nice partners. *The American Lawyer* 12(4), 40.

Morgan, H., and Soden, J. (1979) Understanding MIS failures. *Database* 5, 157–171. [Cited in: Belassi, W., and Tukel, O.I. (1996) A new framework for determining critical success/failure factors in projects. *International Journal of Project Management* 14(3), 141–151.]

Mustafa, A. (1999) Public private partnership: an alternative institutional model for implementing the private financial initiative in the provision of transport infrastructure. *Journal of Project Finance, Summer* 5, 64–79.

Naoum, S., Walker, G., and Fong, D. (2004) *Critical Success Factors of Project Management International Symposium on Globalisation & Construction CIB*. Asian Institute of Technology, Thailand, 17 to 19 November.

National Audit Office (2003) *The Operational Performance of PFI Prisons*. Report of Comptroller and Auditor General, HC 700, Session 2002–2003, The Stationery Office, London.

National Audit Office (2005a) *Darent Valley Hospital: the PFI Contract in Action*. Report of Comptroller and Auditor General, HC 209, Session 2004–2005. The Stationery Office, London.

National Audit Office (2005b) *Improving Public Services through Better Construction*. Report by the Controller and Auditor General. HMSO, United Kingdom Official Publications Database, London.

National Audit Office (2006) *The Paddington Health Campus Scheme*. Report by the Controller and Auditor General, HC 1045 Session. HMSO, United Kingdom Official Publications Database, London.

National Health Service (1999) *Public Private Partnerships in the National Health Service: The Private Financial Service: Good Practice*. HMSO, United Kingdom Official Publications Database, London.

Office of Government Commerce (2002) *Good Practice Guide: Learning Lessons from Post-Project Evaluation*. OGC, London.

Office of Government Commerce (2003) *Achieving Excellence in Procurement Guide 06, Procurement and Contract Strategies*. OGC, London.

Office of Government Commerce (2004a) *Achieving Excellence in Construction, Procurement Guide 09 Design Quality*. OGC, London.

Office of Government Commerce (2004b) *Improving Standards of Design in the Procurement of Public Buildings*. CABE, London.

Office of Government Commerce (2004c) *The OGC Gateway Process*. OGC, London. Available online at http://www.ogc.gov.uk.

Office of Government Commerce (2006) Governance. Available online at http://www.ogc.gov.uk/delivery_lifecycle_governance.asp.

Office of Government Commerce (2008) Governance. Available online at http://www.ogc.gov.uk/delivery_lifecycle_governance.asp.

Office of Government Commerce (2009) *The OGC Gateway Process*. OGC, London. Available online at http://www.ogc.gov.uk.

Office of Public Services Review (2003) *Improving Programme and Project Delivery*. OPSR, London.

Partnerships UK (2006) *Report on Operational PFI Projects*. Available online at http://www.partnershipsuk.org.uk (Accessed 23 March 2006).

Pinto, J.K., and Kharbanda, O.P. (1995) *Successful Project Managers: Leading Your Team to Success*. Routledge, New York.

Pinto, J.K., and Slevin, D.P. (1989) Critical success factors in R&D projects. *Research Technology Management*, January–February, 31–35.

Powl, A., and Skitmore, M. (2005) Factors hindering the performance of construction project managers. *Construction Innovation* 5, 41–51.

Ramo, H. (2002) Doing things right and doing the right things time and timing in projects. *International Journal of Project Management* 20, 569–574.

Robinson, H.S., Carillo, P.M., Anumba, C. J., and Al-Ghassani, A.M. (2005) Knowledge management practices in large construction organisations. *Engineering, Construction and Architectural Management* 12(5), 431–445.

Robinson, H.S., and Scott, J. (2009) Service delivery and performance monitoring in PFI/PPP projects. *Construction Management and Economics* 27(2), 181–197.

Rogers, M. (2001) *Engineering Project Appraisal*. Blackwell, Oxford.

Scott, J., and Robinson, H. (2008) Payment mechanism in operational PFI projects. In: Akintoye, A., and Beck, M. (eds). *Policy, Management and Finance for Public Private Partnerships*. Wiley-Blackwell Publishing, Oxford.

Tukel, O.I., and Rom, W.O. (1995) *Analysis of the Characteristics of Projects in Diverse Industries*. Working Paper. Cleveland State University, Cleveland, OH.

Walker, G., and Naoum, S. (1997) *Determinant of Project Management Success*. Proceedings of Association of Research in Construction Management Symposium. On Organisation and Management. Liverpool.

Westerveld, E. (2003) The project excellence model: linking success criteria and critical success factors. *International Journal of Project Management* 21, 411–418.

Williamson, O.E. (1996) *The Mechanisms of Governance*. Oxford University Press, New York, Oxford.

Winch, G.M. (2001) Governing the project process: a conceptual framework. *Construction Management and Economics* 19, 799–808.

Whelan, E. (2005) Personal Communication on Reporting Structure for a £25 million NHS Build Schemes (Procure 21 Projects). North West London Strategic Health Authority, London.

5

Case Studies on Governance in the Health Sector

The principles of governance, its components and its application in PFI/PPP projects have been examined in the previous chapter. This chapter uses case studies from the health sector to provide evidence on the state of governance and its impact on the delivery of National Health Service (NHS) build programmes in the United Kingdom. The NHS is a complex organisation employing one in twenty of the working population of the United Kingdom. It has an annual expenditure of £42.5 billion and is responsible for over 400 Health Care Trusts whose principal aim is to provide local health care services through 1200 hospital sites and over 11 000 GP surgeries across the United Kingdom (Priestly, 2000). The build facilities through which health care is delivered range from small primary care practices to large, multidisciplinary and specialised hospital sites having a replacement value of over £72 billion (DoH, 2000). Total investment in public services is approximately £50 billion in 2005/2006, compared with £23 billion in 1997/1998 (HM Treasury, 2006), and nearly a third of all investment in the public sector as a whole over a 2-year period between 1999 and 2002 was speculated, by the Treasury, to be privately financed (HM Treasury, 2000). PFI represents one option for infrastructure and facilities investment that enables the government to secure value for money for the extra investment it undertakes (HM Treasury, 2006).

This chapter focuses on some practical aspects of governance in the planning and development phase of PFI/PPP projects. It starts with a brief overview of the health sector showing how the NHS Trust responsible for procuring hospitals interacts with other public bodies and agencies such as the Department of Health (DoH), Strategic Health Authorities, the Treasury and Office of Government Commerce in delivering PFI/PPP projects and other build programmes. The evolution and development of PFI/PPP projects in the health sector are then outlined in the context of recent changes to facilitate the delivery of PFI/PPP projects. Using 'live' capital investment projects of varying sizes and complexity funded through PFI, some aspects or components of governance such as reporting structures and levels of responsibility, project controls and risk management procedures focusing at the planning and development phase are examined to determine the degree to which these were considered by Project Directors as contributing factors to the project delivery outcome. A multiple case study approach was adopted based on semi-structured and open-ended interviews with suitable and accountable persons in each organisation. The outcome of the interviews from Project Directors (considered experts in this subject area) and senior personnel within the organisations formed the basis of a detailed analysis and discussion.

5.2   Overview of Health Sector and Evolution of PFI/PPP Projects

The Department of Health in England and the ministerial equivalent in Scotland (Scottish Government for Health and Wellbeing) and Wales (Welsh Assembly for Health and Social Care) are ultimately accountable for health spending and are responsible for granting or withholding approval of the outline business case[1] for complex NHS build schemes. The health service provision is delegated to the Strategic Health Authorities (England), Regional Health Boards (Scotland) or Local Health Boards (Wales), respectively, through a rigorous process of performance management, and for PFI/PPP projects give approval to proceed from one planning and development stage to the next. The role of the Private Finance Unit (PFU)[2] is to provide guidance to NHS organisations developing PFI schemes. Another supporting body for PFI/PPP schemes is the Office of Government Commerce (OGC). This is an independent office of HM Treasury, established to help government deliver best value from its spending. The OGC works with central government departments and other public sector organisations to achieve the following: the delivery of value for money from third party spend; delivery of public projects to time, quality and cost; getting the best from government estates, delivering sustainable procurement and operations on the government estate and to support government policy goals. Figure 5.1 gives a brief overview of the relationship between NHS Trust[3] and other public bodies such as the Department of Health (DoH), Strategic Health Authorities (SHA), the Treasury (HMT) and OGC in the delivery of PFI/PPP projects.

The first wave of major PFI hospital developments was launched in 1995 with the requirement for the contract to represent value for money when measured against an equivalent project (the public sector comparator) delivered through public funding (HM Treasury, 1997, revised 2003). There

---

[1] In England, the DoH is more involved with full business case with the exceptions of more complex and politically driven projects; these will also involve the DoH at the OBC stage.

[2] The equivalent in Scotland is the Private Finance and Capital Unit (PFCU).

[3] Diagram in Figure 5.1 refers to NHS Trusts; it should however be noted, as of 1 February 2009, that there are 114 NHS Foundation Trusts in operation across England, and the public body, Monitor, is the independent regulator of these Foundation Trusts. The introduction of NHS Foundation Trusts is the beginning of the devolvement of decision-making from central government control to local organisations and communities so they are more responsive to the needs and wishes of their local people. Both SHA and Monitor will look for evidence that the Community Primary Care Trusts support the patient activity assumptions underpinning any project affordability case.

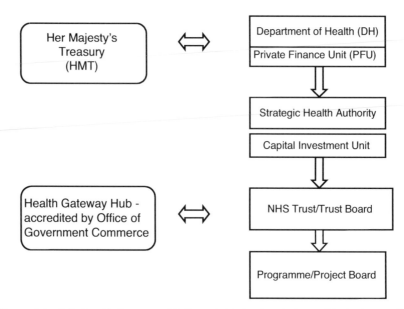

Figure 5.1 Relationship between NHS Trusts in England and other public bodies involved in delivering PFI/PPP projects.

was a delay in the uptake of the PFI approach due to a number of factors and a considerable debate on whether such schemes were providing value for money due to the complex processes involved. Evidence from the first wave NHS schemes suggests that PFI appears to be more expensive than anticipated and has led to enforced reductions in hospital capacity as Trusts have struggled to switch cash from clinical practice to pay for private capital schemes (Gaffney and Pollock, 1999). The UK government recognising the need for improvement appointed Sir Malcolm Bates to review PFI projects, in 1997, which led to a number of recommendations for improvement, among them a Treasury Task Force (TTF) to act as a centre of PFI expertise and more significantly new legislation clarifying the powers of NHS Trusts and other public organisations or authorities to enter into PFI deals (HM Treasury, 1997). Subsequently, a second 'Bates review' of the PFI process led to the largest hospital building programme since the formation of the NHS (DoH, 2000). As a result of the NHS plan in 2000, £7 billion of new capital investment will be procured through PFI, and 40% of the total value of the NHS estate is expected to be less than 15 years old by 2010.

In 2005/2006, 10 years after the launch of the first wave PFI projects, the NHS witnessed a significant amount of new PFI activity. In 2006, six major PFI projects were completed within the NHS, with a further 17 hospitals and other facilities under construction and 45 in the pipeline (DoH, 2006/2007). PFI projects range from relatively small NHS build projects such as the development of an 'imaging centre' costing £25 million to a major hospital development scheme such as The Royal Bartholomew's and London Hospital costing over hundreds of millions pounds, respectively. Whilst the risk associated with the smaller schemes may not be significant, the risks attached to the larger schemes may be fundamental to the ongoing viability of the Trust (London City Audit Consortium, 2005). Each PFI hospital is tendered on a single basis by the appropriate Trust and the actual procuring of public sector schemes through PFI can be a long and convoluted route (Grimsey and Graham, 1997), and more critically, the costs of tendering for PFI are considerably higher than for other procurement systems (Birnie, 1999).

5.3   Case Study Findings on Early PFI Schemes

This section presents the case study findings on two simple early wave PFI schemes (Case Studies 5.1 and 5.2), built on demolished or adjacent brownfield land with a difference in capital costs of approximately £150 million.

Case Study 5.1: Hospital Redevelopment Scheme (< £200 Million)

This case study provides an understanding of the governance and project delivery issues relating to a new hospital development with a capital cost of less than £200 million. The PFI capital project to redevelop the organisation's NHS facilities, dating back to the early nineteenth century, was awarded to a preferred private sector organisation that took responsibility for the design, finance, construction and management of the buildings. The redevelopment involved the demolition of all of the estates oldest buildings and the construction of a brand new building at the centre of the site. This redevelopment was to house state-of-the-art facilities for accident and emergency, intensive care, operating theatres, diagnostic imaging, outpatients and new wards for the serving population.

Governance issues

The Project Board decision to pull-out of the first wave public sector PFI schemes and restart under the second wave provided the opportunity for the PFI project to be reviewed and revaluated in terms of both value for money and affordability. The first step taken by the new appointed Project Director was to introduce a reporting structure that was reflective of the project size and complexity and to ensure that the right stakeholders were being engaged from the onset, such as representatives from the Treasury's PFU and the Department of Health's Capital Investment Unit (CIU). The project was subjected to an early Gateway Review that resulted in a number of basic project management changes but little advice on technical issues such as design development.

One of the early mistakes made on this project was the absence of a project 'working group' that acted as a filter for the Project Board; hence, all work streams reported to the Project Board. This situation resulted in unresolved inter- and intra-working group conflicts that needed to be resolved at board level, thus making the original reporting structure weak and ineffective. Another contributing factor for poor project governance was the absence of key individuals; for example, the project had no dedicated Planning Lead to work with clinical groups making clinical sign-off for output-based specifications (OBS) difficult and delaying the timeline for PFI planning and procurement tender documents. The project also suffered from the introduction of new board members who were not in agreement with the original decisions that were made, and hence, milestones were revisited and equally significant was the suggestion, by the Project Director, that there may have been some wrong appointments to the Project Board, '[W]hen unskilled members of the Project Board are recruited poor decisions are made.'

With regards to levels of responsibilities and approval actions at each milestone, the Project Director was strong in his belief that, while ultimately it was the Chief Executive Officer (CEO), as the Senior Responsible Officer (SRO), who took full project accountability, a collective responsibility through an informed board decision was the preferred way forward. The responsibility to proceed from a key milestone within the planning and development stage such as the OBC to OJEU notice was, in this project, a collective corporate decision. In this 'case study' organisation, the Project Director consistently informed decision-makers at board level through organised workshops that were held prior to each major milestone. Each workshop identified subsequent milestones of the planning and development stage and appropriate advisers would present draft documents and design plans for the group to review and comment. 'Presentation of options and influencing the outcome' is important, said the Project Director, 'to ensure right decisions are being made via an informed process.'

The Project Director was aware of build projects where SRO delegated all the decision-making to the Project Director; this he said 'raised problems where an over-arching corporate decision was important'. The Project Director believed that there is a relationship between roles and responsibilities and project size. He argued that it was good practice to look at project size and complexity and to then build a reporting structure with levels of responsibilities that was capable of supporting the project size and associated complexity. He gave an example of a small project where you would have a simple reporting structure with fewer executives on the decision-making Project Board. In his opinion, the deciding factor for increased levels of responsibilities lies heavily with the basic requirement of the project.

Critical factors and project outcome

The single, most cited factor for project success was project continuity from having the same individuals in the project team for the duration of the project as this ensured key knowledge was retained, managed and transferred from one major milestone to the next. Success at each milestone during the build process of the project was dependent upon effective face-to-face meetings between the PFI Project Co (the vehicle for PFI project delivery), often referred to as the special purpose company (SPC) or vehicle (SPV) and the Hospital Trust, facilitated by the Project Director.

Following a poor start, project performance improved through implementation of a number of risk management systems such as regular review of the risk register and basic project controls such as good document/version control and audit trail of decision-making, assumption monitoring and scenario planning. Surprisingly, the control system that occasionally hindered delivery, in his opinion, was the government-directed external Gateway Review, when members of the review team were not always more knowledgeable about NHS schemes to give credible advice. On such occasion, the reviews were perceived as unnecessary box-ticking exercises.

Initially, there was no contingency planning; the Project Director believed this was due to project managers lacking basic project management skills, and therefore, priority was given for PRINCE2 training. In addition, the absence of effective reporting structure hindered decision-making but again this was quickly rectified following the first Gateway Review and subsequent investment in the project team. Hence, at OBC, the project moved from having low to high levels of project controls. The Project Director also identified the reasons why projects are too often not successfully delivered. This included non-appropriate decision-makers being appointed and wrong decisions being made at a crucial milestone in the project. He argued that having good governors in place at the right point of the project is necessary to avoid poor delivery. The Project Director was quite clear on what aspects of governance ensured successful delivery of project milestones. He argued that the governing framework for successful project delivery is ensuring that individuals have the correct project role and responsibilities that will vary for each stage of the planning and development phase.

In summary, the initial project for this case study organisation was poorly conceived and resulted in the scheme restarting and moving from the first wave of PFI schemes to the second wave. The final project was delivered after a long and hard battle with the private project company (Project Co).

Case Study 5.2: Hospital Redevelopment Scheme (< £400 Million)

The capital project for this case study organisation involved a new hospital development that brought together two hospitals of a single organisation onto a single site. This new build with a capital cost in the region of £300–£400 million was procured under the second wave of NHS PFI procurement. The 1250-bed new hospital is considered to be one of Europe's largest and high-technology hospitals having equipped itself with some of the latest medical equipment including the gold standard diagnostic scanners for early diagnosis of cancer and Picture Archiving and Communication Systems (PACS) that produce x-ray results at the touch of a screen. The hospital also has 27 operating theatres with capital replacement plans for all the major items of equipment in servicing departments to be replaced every 7–10 years.

Governance issues

The Project Director for this case study organisation stated the reporting structure as a critical component of governance. The top–down structure was described as having several work streams at the ground level reporting to a project team (headed by a project manager), which reported to an oversight Project Board. The project team skill-mix changed as the project progressed from business case through to selection of preferred bidder and construction. The Oversight Project Board was considered important, and in particular, the membership which included executive members, construction lead, designer, facilities manager and top management members from Project Co (the vehicle for PFI project delivery). Right membership allowed project concerns from three different viewpoints (i.e. NHS client, Construction company and Project Co or SPC/SPV) to be heard and actioned immediately. Particularly important was the relationship between the Hospital Trust and the construction team, designers and facilities management (FM) staff. The Project Director also recognised that the bigger the project, the more complex the decision-making process, as it involves more people.

Failure to maintain an audit trail of decision-making at key milestone points, such as signing off 1:50 room data sheets was a major cause of project delays. The project team was completely replaced by a new team 3 months after the start of construction, but fortunately, the Project Director saw this 'change in personnel' as an important governance issue and ensured that the programme had good records and audit trail of change request. Planning responsibilities, regardless of project stage, firmly stood with clinical leads and service managers, and to ensure ownership of service redesign, the clinical leads reported to the project team for two-way communication and direction.

Critical factors and project outcome

Although there was a change in the original project team, the overriding critical success factor for this project was a disciplined team, which understood what had to be done, when and by whom. Again, successful delivery of services was reported to be through having regular workshops for all stakeholders; regular updated programme charts by project managers and planning tools such as process mapping. Project governance was maintained through an integrated process of controls and risk management systems. The only hindrance was an ill-designed PFI contract that lacked corporate participation. The Project Director believed governance in the project helped to bring about discipline, which was viewed as a key success factor.

The following factors were promoted, by the Project Director, as key to the successful delivery of this new hospital: A simple but robust reporting structure that was right for the project at hand; the correct representatives in the committee; the right committee involved at the different levels; the correct work being performed at the right levels and clear lines of communication both vertically and horizontally. In this particular project, the Liaison Committee was given the credit for good communication strategy bringing together representatives of the project (i.e. the Trust, construction team, design team, FM staff and Project Co).

High levels of project controls were maintained through simple and effective reporting structure with clear level of responsibilities. Also effective was the use of PRINCE 2 project management by skilled project managers and effective communication strategy for managing both internal and external stakeholders.

The Project Director achieved the delivery of the new hospital on time and within resources allocated despite inheriting the scheme at financial close when decisions on designs were already frozen at 1:200 scale drawings. After a period of 4.5 years, the new University Hospital opened to provide local people with one of the best hospitals in Europe on their doorstep.

5.4   Case Study Findings on Complex PFI Schemes

The case studies reflect the findings of two highly publicised and ambitious NHS PFI Build Projects that were heavily influenced by both internal and external governing politics prevailing at the time. Equally, both organisations reported a constant battle against clinical resistance to 'change'. The status of governance and its impact on the build programmes adopted by each case study organisation are presented below.

Case Study 5.3: Multi-Trusts PFI Scheme Co-Located on a Single Site

This case study involves the development of a large PFI scheme that requires significant investment in the provision of local and specialist health care services. The business case identified a single site as the preferred option for redeveloping services currently provided by three organisations. The proposed scheme embodied a vision for the wholesale redevelopment and renewal of services and facilities that are currently dispersed across several sites with outmoded accommodation. The objective was to create a solution for patients, both locally and nationally, with modern health care and research facilities incorporating latest technologies.

This case study organisation represents more than one NHS Trust co-locating on a single site together with a third non-NHS Organisation (University). The project was complex for three reasons: firstly, there were three organisations involved and therefore three CEO/SRO with often conflicting priorities. Secondly, the non-NHS organisation had slightly different corporate processes for financial accounting, and thirdly, the two NHS Trusts were to remain as separate entities. The Project Director appointed in the later years was an external candidate with no previous associations with any of the three organisations and had good knowledge and experience of PPP projects, having previously managed a new build PFI Hospital. Case Study 5.3 focuses on governance issues in the early planning and development phase of this complex multi-organisation PFI scheme, involving co-location.

Governance issues

The Project Director was appointed to this project for his knowledge and experience from having recently commissioned a new PFI Hospital. He described governance as 'a decision and responsibility hierarchy that is fit for purpose, without which there will be no successful projects'. He then identified the ingredients of good governance as 'a structured hierarchy with direct access to the Trust Board, plus experienced people with the right skills in the right place in the reporting structure'. In his view, having good project governance is important and he qualified this by adding, '[T]he more rigorous the governance in the early stages of the project the better is downstream delivery of the project.' This was re-emphasised when he argued that the ingredient lacking from this project was lack of investment in governance at the front end of the project, that is at the OBC stage.

However, the early decision-making process during the period of project conception (1997/1998) was somewhat flawed with Senior Managers from one of the organisations leading the project with some support from the partnering organisations. The process improved with the introduction of an additional decision-making tier in the form of an overarching group with executive members from all three organisations to oversee the work of the Joint Project Board. The decision-makers on the Joint Project Board were generally of a wide skill-mix but lacked individuals with specific PFI and project management experience for a project of this size and complexity. The Project Director recognised that levels of responsibilities vary depending on the chosen procurement strategy, project size and complexity. He noted that projects, particularly of this size, require more experienced people with greater levels of responsibilities whereas smaller projects can get away with fewer, less experienced people. Hence, for a project of this size, he argued that there was potential for more project risks.

In terms of management systems, the project was controlled via a complex matrix of project programming linking timeline charts, critical path, progress monitoring, action lists and update reviews. Poor project delivery of early milestones were cited to be mainly due to ill-defined project start-up, inadequate front-end resources, lack of relevant experience in design and technical working group and poor communication between the Project Board and key stakeholders such as the commissioners. These issues were constantly being addressed, and during the latter years of the project, there was evidence of a strong project team developing through appointments of internal and external advisers with hands-on experience of projects.

Critical factors and project outcome

Both the Project Director and the Senior Manager interviewed agreed that during the conception stage when in-house managers were managing the project, there was a definite failure in the process of decision-making and even those decisions made were not effectively carried out. The reason cited for this was the absence of a wide range of stakeholders at the board meeting and therefore decisions took longer than normal through this absence of critical decision-makers being present at all times. Two years later, when systems were eventually put in place, decision-making improved with the right people at the right time but consensus on critical issues still took longer than normal as clinical opinion for the two Trusts were often opposing due to conflicting priorities.

The Project Director and Senior Manager believed that once a more robust reporting system was embedded, the decisions emerging were correct but politics influenced decisions overall and either one or two of the Trust Boards often wavered on decisions previously made by the Joint Project Board. Lack of knowledge in specific areas of PFI and lack of project management experience for a project of this size hindered decision-making. The reporting structure was believed to be effective in the latter years but was introduced too late for the early planning stages.

The key success factors identified for project delivery were consistent leadership and development of an effective strong cross-Trust project team committed in their belief of achieving the long-term benefits of the project. Unfortunately, the success factors were not fully realised as the project team constantly battled to keep parity in planning actions between the two Trusts and University. Hence, the project suffered in terms of consistent and timely delivery of milestones despite having robust project management systems in place to support the project such as good programming with timeline charts, critical path planning with aligned progress monitoring and regular reviews, and action lists to support programme and risks analysis/registers with review updates at key stages. There was also a problem with overzealous resource scheduling and information monitoring that was too difficult to control and therefore required more external resources to manage effectively. Furthermore, the level of control was considered low, as these controls were not adequately resourced to make them effective for a project of this size.

Case Study 5.3, during a period of political uncertainties and following some costly reiterations of the new build options, was terminated after a second rewrite of the OBC. The outcome was a failed scheme due to many non-deliverables.

Case Study 5.4: Multi-Site PFI Scheme

The organisation involved in Case Study 5.4 represents another complex project spanning years of public consultation and eventual merging of two separate Trusts with pioneering medical history and logistics into one single Trust. This scheme started during the early entry of PFI into the public health sector but the project's complexity and strong public opinion against the closure of one of the two hospital sites led to various reiterations of the original scheme. The strategic direction for a two-site solution, rebuilding two large major district hospitals, was confirmed by the Turnberg Review (1997) and endorsed by the government. The Project Director interviewed took the project from OBC to the completion of the full business case.

Governance issues

The Project Director considered governance as important for three reasons: for Trust-side decision-making, project protection and giving confidence to bidders in terms of the project's viability. The project, when inherited at OBC stage, lacked any readily available guidance from the Treasury on project management of major capital projects and hence lacked the reporting structure required for a 'New Hospitals Project Board' and an in-house project group for managing the many new hospitals projects. It was clear that governance was not well articulated at project conception or at the early OBC stage. The Project Director noted, '[T]here was no assurance of how the project was proceeding until the FBC stage, when the project was subjected to the government directed but external, Gateway Review.' The Gateway Review team consisted of health professionals with experience of similar project management issues. The Project Director further argued that governance was important in an alteration or new-build PFI/PPP scheme to support the transformational change in the way in which health care is delivered by the Trust in the future. He noted that this importance is reflected in the guidance now made available from the OGC but there is still the issue that it will take some time for this knowledge to filter into the system and become common practice. During FBC stage with more government-directed guidance available to the Project Director, it is clear that governance of the project had improved a lot. The project at the time of the interview was said to have benefited from having an informed board and CEO with a robust framework for decision-making.

Poor governing structure or processes during the early days was, as with Case Study 5.3, narrowed down to lack of experienced people in the project team and board which meant there was absence of clear direction from the board. Project governance dramatically altered once past the OBC stage, with restructuring of the reporting structure following the government-directed Gateway Review. Poor project discipline and direction was immediately addressed with the formation of a transparent and robust decision-making structure. This further improved as 'good practice' guidance from the government websites started to filter into the system and become common practice. By the time the project reached FBC status, the board was made up of more experienced and knowledgeable non-executives who challenged proposals and gave the project team clearer actions to take forward with confidence.

Critical factors and project outcome

Although there were problems at the beginning, the project clearly benefited from two key factors, the peer review by the Gateway team and the application of project management methodology, PRINCE 2. The Project Director explained that there was no requirement for a SOC and accordingly no structured allocation of roles and responsibilities was made early in the project. The first project stage was therefore the completion of the OBC. Whilst the board initially approved this, there was still no formal reporting structure and boundaries of authority were unclear. In the absence of a formal reporting structure, the project lacked discipline and presence of authority to ensure consistency in the decision-making process. This was confirmed by the findings from the Gateway Review team, who reported the absence of a formal reporting structure, resulting in a number of projects moving off at a tangent from the core objectives. This message from the Gateway Review resulted in the formation of a Project Board (New Hospitals Board) to govern the entire capital programme for the Trust. The reason given for poor project governance initially was quite simply ignorance and lack of understanding of its importance by the wider executive team which meant there was no top–down drive for good governance.

The Project Director felt that governance had not played a big enough role in this particular project, although when embedded at a later stage, it was found to be most helpful. He cited difficulties that could have been avoided with good governance. For example, with regards to decant issues, good governance could have alerted the team to dangers ahead much earlier and set appropriate contingency actions in place.

The levels of control in this project moved from initial low levels to high levels following recommendations by the Gateway Review team. Hence, subsequent to Gateway 1, all project managers were trained to use PRINCE 2, project management tools. The absence of effective controls and project management in the early stages of the project contributed to delays and overspending on advisers; however, levels of control in this project moved from initial low levels of project control to high levels following implementation of recommendations by the Gateway Review team at Gateway 1 and 2 (GW1 and GW2).

Case Study 5.4, a large and complex project with a high degree of political interest and pressure to deliver flagship 'Super Hospitals', completed financial close and took the next step of moving into the construction phase and preparing operationally to realise its benefits from creating a centre of health care excellence.

5.5   Analysis and Discussion of Case Studies (5.1–5.4)

This section analyses the case studies presented in the previous sections and discusses key themes of governance affecting project delivery. The findings in each organisation's approach to project governance and the relationship with project delivery are compared and summarised in Tables 5.1–5.7.

5.5.1   Reporting structure and levels of responsibilities

Case Studies 5.1, 5.2 and 5.4 had simple reporting structures that are reflective of the project management methodology, PRINCE 2, endorsed by the Department of Health for the NHS (DoH, 1994; CCTA, 1996). In Case Study 5.3, with the absence of a single responsible SRO, the reporting structure was visible but complex. Table 5.1 is a summary of issues relating to the reporting structure and levels of responsibilities.

In projects where roles and responsibilities are not clear (e.g. Case Study 5.3) or reporting structures are ineffective (e.g. Case Study 5.1) or absent all together (e.g. in the initial stages of Case Study 5.4), the chances of project failure is arguably high (DoH, 1994; CCTA, 1996; NAO, 2006).

5.5.2   Effective controls

Inadequate controls are cited in project reviews (NAO, 2006) as the leading source of governance problems. Capital projects need rigorous processes

Table 5.1   Reporting structure and responsibilities.

	Case Study 5.1 (< £200 million)	Case Study 5.2 (< £400 million)	Case Study 5.3 (> £500 million)	Case Study 5.4 (> £500 million)
Reporting structure	Simple but initially ineffective due to poor reporting lines; all work streams reported to the Project Board	Simple and effective	Complex: Three Senior Reporting Officers and two Trust Boards.	Reporting structure absent in the initial stages of the project. Robust and simple structure established during the OBC stage
Levels of responsibility	Level 1: Project work streams Level 2: Project Board Level 3: DoH Key responsibility is with Chief Executive as the SRO	Level 1: Project Working Groups Level 2: Project Director Level 3: Project Board Level 4: DoH Key responsibility is with Chief Executive as the SRO	Level 1: Project Manager; Head of Planning and Head of Design & Technical Team Level 2: Project Director Level 3: Project Board and SRO. Level 4: Trust Boards Level 5: DoH	Level 1: Project Manager Level 2: Project Director Level 3: Project Board Level 4: Trust Board Level 5: DoH

and controls to ensure their delivery aligns with the organisation's overall strategic goals. Table 5.2 is a summary of the project controls and levels of controls associated with each case study organisation. All four case study organisations indicated their awareness of the importance of having in place robust and effective project controls but did not necessarily implement the best approach, particularly at project start (e.g. Case Studies 5.1 and 5.4, see Table 5.2) or adequately for a project of great magnitude in size (e.g. Case Study 5.3, see Table 5.2).

A number of tools are used by the case study organisations. Jaselskis and Ashley (1991) suggested that by using project management tools such as milestone delivery charts, the project managers would be able to plan and execute their construction projects to maximise their chances of success. The one case study organisation that clearly excelled in the effective use of management tools to drive management actions was Case Study 5.2. Project control in Case Study 5.1 had too much reliance on the Project Director and was not, therefore, a robust system. Case Study 5.3 lacked effective project control quoting complexity of the project as the underlying issue. However, Case Study 5.4, which was equivalent in size and complexity, demonstrated project effectiveness through the use of a simple control system. Subsequent findings on project outcome (Section 5.7) reveal a degree of correlation between levels of controls and successful project delivery.

Table 5.2   Project controls and levels of control.

	Case Study 5.1 (< £200 million)	Case Study 5.2 (< £400 million)	Case Study 5.3 (> £500 million)	Case Study 5.4 (> £500 million)
Project controls	Basic project controls managed by Project Director No contingency planning initially	Excellent project control delivered via effective use of management tools and discipline	Complex controls were not adequately resourced for effectiveness in a project of this size High levels of political stakeholder influence	Clear and visible project controls implemented following the first Gateway Review High levels of political stakeholder influence
Levels of control	Moved from low to high levels of project controls	High levels of project controls maintained from project SOC to project handover (New Hospital Operational)	Complex controls but at low level as it is not effectively managed	Moved from low to high levels of project controls

5.5.3   Project management

'In a technology driven environment where organisational change and developments are becoming increasingly important, tools such as project management can provide a useful way for organisations to manage that change effectively' (Munns and Bjeirmi, 1996). Table 5.3 shows the project management tools employed by the case study organisations and its effectiveness measured in terms of project performance. Assessment of project performance was subjective based on interview with Project Directors and Senior Managers. Performance rating of the project is measured on a scale of 1 (low performance) to 5 (high performance). The project management methodology endorsed by the DoH for the NHS is PRINCE2; this was used in all four case study organisations. However, in Case Study 5.3, whilst application of PRINCE2 is evident, the absence of knowledge and skills for a project of this size and complexity resulted in poor project management and communication (Table 5.3).

5.5.4   Risk management

Although a certain amount of risk is unavoidable, the public expects government agencies to identify and manage that risk systematically to ensure that the project remains under control (NAO, 2006). Table 5.4 shows the risk management issues and level of risks associated with the projects in each case study organisation. Case Study 5.1 clearly lacked a robust risk management

Table 5.3   Effectiveness of project management.

	Case Study 5.1 (< £200 million)	Case Study 5.2 (< £400 million)	Case Study 5.3 (> £500 million)	Case Study 5.4 (> £500 million)
Project management	Project managed by Project Director using PRINCE2 methodology

Good communication process | Robust management systems (PRINCE2)

Good performing project team led by a strong Project Director both during OBC and construction

Good communication | Project management (PRINCE2) evident but absence of knowledge and skills for a project of this size and complexity resulted in poor project management structure and poor communication between stakeholders | Project management improved with good delivery at late OBC and FBC stage

Post Gateway Review, all project managers were trained in PRINCE2 |
| Performance rating (1 – low – to 5 – high) | 3 out of 5 (+++) moving up to 4 out of 5 (++++) | 5 out of 5 (+++++) | 3 out of 5 (+++) | 3 out of 5 (+++) Moving up to 4 out of 5 (++++) |

system that ensured staff take ownership of planned organisational changes. Hence, Case Study 5.1 was left open to the possible risk of design changes occurring as and when clinical teams were enlisted for consultation affecting their departments. The project with the greatest risk of failing is Case Study 5.3, not only because of its size and number of external stakeholders

Table 5.4   Risk management issues.

	Case Study 5.1 (< £200 million)	Case Study 5.2 (< £400 million)	Case Study 5.3 (> £500 million)	Case Study 5.4 (> £500 million)
Risk management	All work streams reported to the Project Board

No project manager at ground level to review day-to-day issues | Project risks managed at front-line service level by clinical directors and, therefore, good management of all organisational risks | Absence of a single SRO put the management of the project at high risk from conception

Risk register not updated | Risk management improved with experienced and knowledgeable recruitment of non-executives to the Project Board |
| Levels of project risk | Medium-risk project | Low-risk project | High-risk project | Moved from high- to low-risk project |

but also because of the absence of a single SRO to champion the project in its entirety (Table 5.4). Tukel and Rom (1995) noted that one of the most critical factors for the successful completion of projects is top management support and the support is usually strongest if there is a project champion from the top management team. The highest level of governance for any NHS PFI scheme is assigned to the SRO (DoH, 2006/2007). For Case Study 5.3, this was somewhat confusing as there were three SRO resulting in poorly managed complex controls and risk management systems. The project was arguably a high-risk project from conception given the number of different stakeholders and interests.

Mott MacDonald (2002) in his government-directed review of large public sector projects concluded that there was a strong correlation between project size and the number of project risks. Both the projects in Case Studies 5.3 and 5.4 were initially at high risk, a view that is shared by the Project Director of Case Study 5.3 who argued that large project size was the reason for increase in number of inherent project risks (Table 5.4). However, the level of risks was subsequently reduced in Case Study 5.4 through the effective recruitment of experienced and knowledgeable Project Board executives. Akintoye *et al.* (2003) demonstrated that successful delivery of schemes depends, amongst other factors, on detailed risk analysis and appropriate risk allocation. Since project risks are inevitable, the management of risks must be optimised and not ignored, and strategies for coping with these risks are needed (Akintoye *et al.*, 2002).

5.5.5   Critical success factors (CSF) in projects

Rockart (1982, 1986) defines 'critical success factors (CSF) as those few key areas of activity in which favourable results are absolutely necessary for a manager to reach his/her goals'. For the case study organisations, the key areas of activity were identified as establishing a case for change (i.e. strategic outline case), developing a value-for-money business case (i.e. OBC), procurement to select a PFI partner, developing a full business case with PFI partner (FBC), contract negotiation, construction and post-project evaluation. More recently, Qiao *et al.* (2001) established eight independent CSF in build, operate and transfer projects that included stable political and economic situation, reasonable risk allocation, selection of suitable subcontractors and management controls. Table 5.5 shows the CSF associated with the case study organisations. Case study organisations 5.3 and 5.4 were subjected to greater political and economic risks than 5.1 and 5.2 for the simple reason that both were requesting significantly high levels of funding (greater than £500 million) as part of the government's objective to create 'Super Hospitals'.

PFI/PPP projects are increasingly used in the United Kingdom to improve public facilities and services provision despite some early bad publicity (Birnie, 1999). A number of factors correlating to the success of PFI/PPP projects have been identified (Keene, 1998; Qiao *et al.*, 2001; Li *et al.*, 2005). However, their importance relative to one another has received less

Table 5.5 CSF for project delivery.

	Case Study 5.1 (< £200 million)	Case Study 5.2 (< £400 million)	Case Study 5.3 (> £500 million)	Case Study 5.4 (> £500 million)
Factors contributing to successful delivery of key milestones	Good project management following a poor start Clear project objectives Good communication between contractors and design team	Good project management and investment decisions Strong leadership for project team Clear roles and responsibilities Good communication between contractors and design team	All contributing control factors appeared to be inadequately resourced to be 100% effective for a project of this size Some good communication between key stakeholders	Effective project monitoring/control Knowledgeable Project Board members making investment decisions Some good communication between key stakeholders

attention until recently when Li *et al.* (2005) collated the CSF and systematically ranked them, particularly in terms of the attention that should be given to them in the development stages of projects (i.e. planning and development phase). They identified four high-loading CSF components, one of which is good governance. Good governance is identified by Frilet (1997) and Badshah (1998) as a CSF in PFI projects. Similarly, all four Project Directors in the case study organisations identified project governance as critical to project delivery. This included Case Study 5.3, although the governance components, whilst all in place, were not adequately supported or managed for a project of such size and complexity.

Other CSF in PPP/PFI projects include 'soft issues' such as social support (Frilet, 1997), commitment (Stonehouse *et al.*, 1996), communication (Pinto and Slevin, 1989), characteristic of project team leader (Pinto and Slevin, 1989) and mutual benefit (Stonehouse *et al.*, 1996). All case studies had identified the soft issue of communication, a key aspect of governance, as an important success factor for project delivery. Case Study 5.2 also identified strong leadership critical, in the organisational and project structure, a key component of governance. To summarise, the Project Directors of the four case study organisations identified key governance factors contributing to successful delivery as good project management, clear project objectives, strong leadership, clear roles and lines of responsibilities, knowledgeable decision-makers, good project monitoring and effective controls.

5.5.6   Critical failure factors in projects

Rubin and Seeling (1967) are one of the first researchers to introduce success and failure factors in projects. They investigated the impact of a project manager's experience on the project's success or failure and concluded that

Table 5.6 Critical failure factors to project delivery.

	Case Study 5.1 (< £200 million)	Case Study 5.2 (< £400 million)	Case Study 5.3 (> £500 million)	Case Study 5.4 (> £500 million)
Factors contributing to poor delivery of milestones	Poor governors in place at critical points in the project Unclear roles and responsibilities	Disbanding of the original project team at the point of construction	Shifting leadership Lack of experienced and knowledgeable staff Conflicting objectives by organisations involved Limited buy-in from stakeholders Risk register not updated	Inadequate resources applied at the front end of project planning Unrealistic and manageable timescale Lack of contingency planning Some poor decisions at project conception

the size of the project managed does affect the manager's performance. This finding reflects the difficulties experienced by the two Project Directors managing and controlling the large and complex projects covered by Case Studies 5.3 and 5.4. Both Project Directors commented on the size of the project requiring more resources particularly at the front end of the project and the need to employ good advisers with good track record of managing significant projects. Avots (1969) identified three main reasons for project failure: the wrong choice of project manager, the unplanned project termination and unsupportive top management. Inadequate executive and ministerial support were contributing factors towards the final decision to abort the project in Case Study 5.3. Table 5.6 shows the critical failure factors associated with the case study organisations.

Project Director of Case Study 5.1 identified poor governors and project team's unclear roles and responsibilities as the main reasons for poor project delivery at key milestones. The case study organisations with relatively larger and more complex PFI schemes (Case Studies 5.3 and 5.4) appear to report a much longer list of factors contributing to non-successful delivery of key milestones. Most of these factors can be classified as aspects or components of governance (Table 5.6).

5.5.7   Project outcome

'Many factors of success can be linked to components of governance such as approval process, procurement process, project controls, project accountability and risk management' (Williamson, 1996; OGC, 2009) to support key decisions in the planning and development phase. The failure factors

Table 5.7   Key factors affecting project outcome.

	Case Study 5.1 (< £200 million)	Case Study 5.2 (< £400 million)	Case Study 5.3 (> £500 million)	Case Study 5.4 (> £500 million)
Key factors affecting project delivery	Having a good communication strategy and a simple reporting structure post-OBC stage were key reasons for final project success	Robust reporting structure, clear lines of responsibility and communication were the key to successful project delivery	Lack of resources and experienced project team for a project of this size plus, poor communication between organisations all contributed to poor delivery at key milestones resulting in delays and reiteration of OBC.	Ineffective controls and project management in the early planning stages led to project delays and overspending on advisory budget Knowledgeable decision-makers at the later stages of planning led to successful delivery of OBC and FBC
Delivery outcome and project status	Post-construction, successful operation and service delivery of new hospital to public but over-budget and overtime	Post-construction, successful operation and service delivery of new hospital to public within budget and within time	Many non-deliverables (failed scheme). Project terminated in the planning and development phase of the project; the SHA did not approve the OBC halted all work in preparation for Invitation to Negotiate (ITN)	OBC and FBC delivered successfully but construction delayed and over-budget Project now under construction

recorded by the case study organisations such as unclear roles and responsibilities, poor reporting structure, weak leadership, poor controls and project management can also be addressed by implementing a governance framework. Table 5.7 shows the key factors affecting project delivery (success or failure) in each case study organisation.

Governance of people is achieved by having in place a reporting structure whilst governance of processes is achieved through project controls and risk management processes (Winch and Carr, 2001). The clear success of Case Study 5.2 can be attributed to successful governance of both people and processes, whereas project completion overtime and over-budget in Case Studies 5.1 and 5.4 are a result of poor governance at the initial planning and development phase of the PFI schemes. The delivery outcome of Case Study 5.3, the failed PFI scheme, can be argued to be reflective of the poor governance of people and inadequate process governance for a project of this size and complexity. In summary, factors affecting project delivery can therefore be linked to 'soft' issues relating to people and 'hard' issues relating to processes, which are all essential components of governance.

5.6 Concluding Remarks

The case studies demonstrate that good governance is integral to achieving excellence in delivering PFI/PPP projects through a controlled and managed governance framework. In the absence of an effective organisational structure and robust control and monitoring mechanisms, PFI/PPP projects are in danger of running overtime, incurring increased expenditure and additional cost on external advisers for public sector clients. Furthermore, more complex projects require bespoke governance framework of controls and monitoring mechanisms that reflect organisations leadership, decision-making systems, organisational structure and communication strategy to ensure right project deliverables for the right project. Organisations must therefore ensure that their governance framework is in alignment with both project size and complexity to achieve project success.

A further implication is that whilst governance tools to support NHS PFI/PPP projects are readily available to the project manager, these are of little value without the provision of appropriate training as to 'when' and 'how' to use these tools effectively. Failure to provide adequate training on governance tools could result in situations where the tools are labelled as 'academic' and fall into disuse. The following are components of governance identified by the case study organisations as CSF for delivering complex NHS PFI build schemes:

- Robust reporting structure
- Clearly defined responsibilities
- Strong leadership
- Good project management skills
- Good communication strategy
- Effective project controls and tools such as risk management systems

Recommendations following the case study research are that public sector organisations should consider the following to successfully deliver PPP/PFI projects:

(a) To develop a governance framework for the build project that is aligned to the wider organisational and project objectives. Within the governance framework, there is a need to ensure clarity in the organisational structure, relationship between different organisational and project units and between a person's 'role', team roles and their associated 'responsibilities'.

(b) To understand the relationship between the key components of governance, organisational structure (e.g. people, teams, roles, coordination, communication strategy) and control and monitoring tools (e.g. processes, standards, tools and project compliance), to facilitate continuous improvement in project delivery.

(c) To make the best opportunity at critical stages of the planning and development phase, in particular the peer review of the project such as that provided by the independent Gateway Review team. Governance at the planning and development phase is the most crucial as mistakes made at the early stage are costly and can have major consequences (as in Case Study 5.3), which can also impact on the construction phase as well as the service delivery and operational phase.

References

Akintoye, A., Beck, M., Hardcastle, C., Chinyio, E., and Asenova, D. (2002) *Framework for Risk Assessment and Management of Private Finance Initiative Projects*. Final Report EPSRC/DTI. Glasgow Caledonian University.

Akintoye, A., Hardcastle, C., Beck, M., Chinyio, E., and Asenova, D. (2003) Achieving best value in private finance initiative project procurement. *Construction Management and Economics* 21, 461–470.

Avots, I. (1969) Why does project management fail? *California Management Review* 12, 77–82. [Cited in: Belassi, W., and Tukel, O.I. (1996) A new framework for determining critical success/failure factors in projects. *International Journal of Project Management* 14(3), 141–151.]

Badshah, A. (1998) *Good Governance for Environmental Sustainability, Public Private Partnerships for the Urban Environment Programme* (PPPUE). United Nations Development Program, UNDP, New York.

Birnie, J. (1999) Private finance initiative (PFI) – UK construction industry response. *Journal of Construction Procurement* 5, 5–14.

CCTA (1996) *Managing Successful Projects with PRINCE2*. HMSO, United Kingdom Official Publications Database, London.

Department of Health (DoH) (1994) *Capital Investment Manual*. HMSO, United Kingdom Official Publications Database, London.

Department of Health (DoH) (1997) *London Strategic Review Report, Chaired by Sir Leslie Turnberg, Set the Context for the Wider London Healthcare Strategy*. HMSO, United Kingdom Official Publications Database, London.

Department of Health (2000) *The NHS Plan, A Plan for Investment, A Plan for Reform*. HMSO, United Kingdom Official Publications Database, London.

Department of Health (DoH) (2006/2007) *Good Practice Guidance: Public Private Partnership in the National Health Service: The Private Finance Initiative*. Available online at http://www.dh.gov.uk/ProcurementAndProposals/PublicPrivate-Partnership/PrivateFinanceInitiative/PFIGuidance/fs/en.

Frilet, M. (1997) Some universal issues in BOT projects for public infrastructure. *International Construction Law Review* 14(4), 499–512.

Gaffney, D., and Pollock, A. (1999) Pump priming the PFI: why are privately financed hospital schemes being subsidised. *Public Money and Management* 17(3), 11–16.

Grimsey, D., and Graham, R. (1997) PFI in the NHS. *Engineering Construction and Architectural Management* 4/3, 215–231.

HM Treasury (1997) *Review of the PFI by Sir Malcolm Bates, Summary and Conclusion*. HMSO, United Kingdom Official Publications Database, London.

HM Treasury (1997, revised 2003) *Appraisal in Evaluation of Central Government, 'The Green Book'*. HM Treasury, London.

HM Treasury (2000) *Public Private Partnerships – The Governments Approach*. HM Treasury, London.

HM Treasury (2006) *PFI: Strengthening Long-Term Partnerships*. HM Treasury, London.

Jaselskis, E.J., and Ashley, D.B. (1991) Optimal allocation of project management resources for achieving success. *Journal of Construction Engineering Management* 117(2), 321–340.

Keene, W.O. (1998) Re-engineering public–private partnerships through shared-interest ventures. *The Financier* 5(2 and 3), 55–59.

Li, B., Akintoye, A., Edwards, P.J., and Hardcastle, C. (2005) Critical success factors for PPP/PFI projects in the UK construction industry. *Construction Management and Economics* 23, 459–471.

London City Audit Consortium (2005) *Guide to the Corporate Governance and Internal Audit of Major NHS PFI Schemes*. HMSO, United Kingdom Official Publications Database, London.

MacDonald, M. (2002) *Review of Large Public Procurement in the UK*. HM Treasury, London.

Munns, A., and Bjeirmi, B. (1996) The role of project management in achieving project success. *International Journal of Project Management* 14(2), 81–87.

National Audit Office (2006) *The Paddington Health Campus Scheme: Report by the Controller and Auditor General, HC 1045 Session*. HMSO, United Kingdom Official Publications Database, London.

Office of Government Commerce (2009) *The OGC Gateway Process*. OGC, London. Available online at http://www.ogc.gov.uk.

Pinto, J.K., and Slevin, D.P. (1989) Critical success factors in R&D projects. *Research Technology Management* 32(1), 31–35.

Priestly, K. (2000) *NHS ProCure21 – Building Better Health Letter*. HMSO, United Kingdom Official Publications Database, London.

Qiao, L., Wang, S.Q., Tiong, R.L.K., and Chan, T.S. (2001) Framework for critical success factors of BOTprojects in China. *Journal of Project Finance* 7(1), 53–61.

Rockart, J.F. (1982) The hanging role of the information systems executive: a critical success factors perspective. *Sloan Management Review* 24(1), 3–13.

Rockart, J.F. (1986) A primer on critical success factors. In: Christine, V.B. (ed.). *The Rise of Managerial Computing: The Best of the Centre for Information Systems research*. Dow Jones-Irwin, Homewood, IL.

Rubin, I. M., and Seeling, W. (1967) Experience as a factor in the selection and performance of project managers. *IEEE Trans Eng Management* 14(3), 131–134. [Cited in: Belassi, W., and Tukel, O.I. (1996) A new framework for determining critical success/failure factors in projects. *International Journal of Project Management* 14(3), 141–151.]

Stonehouse, J.H., Hudson, A.R., and O'Keefe, M.J. (1996) Private public partnerships: the Toronto Hospital experience. *Canadian Business Review* 23(2), 17–20.

Tukel, O.I., and Rom, W.O. (1995) *Analysis of the Characteristics of Projects in Diverse Industries, Working Paper*. Cleveland State University, Cleveland, OH.

Williamson, O.E. (1996) *The Mechanisms of Governance*. Oxford University Press, New York, Oxford.

Winch, G.M., and Carr, B. (2001) Processes maps and protocols: understanding the shape of the construction process. *Construction Management and Economics* 19, 519–531.

6

Knowledge Management in Collaborative Projects

6.1   Introduction

There is an increase in collaborative projects such as partnering, alliances, joint ventures, framework agreements and PPP/PFI projects. PFI/PPP projects are complex, expensive and require long-term commitment between various private sector organisations that have to collaborate and share knowledge to develop solutions that meet the needs of public sector clients. Many organisations have started to use various knowledge management tools such as project extranets to facilitate collaborative working on specific projects. However, 'grafting' knowledge from outside an organisation's environment is challenging and requires mechanisms to bring external and public knowledge to benefit the development of the project. PFI/PPP projects *involve long-term collaboration and* networking across different professional groups such as architects, planners, engineers, surveyors, lawyers, financial specialists and other team members with specific functional relationship in the special purpose vehicle or company (SPV/SPC) to deliver services according to the public sector clients' output specification.

The previous chapters have examined the importance of governance in successfully executing PFI/PPP projects. Knowledge management can play a catalytic role in improving governance through the development of processes and tools, the acceleration of learning to improve the decision-making ability of actors in PFI/PPP projects. This chapter focuses on the theory, principles and application of knowledge management. The concept and different types of knowledge as well as the dynamics of knowledge creation are explained using Nonaka and Takeuchi's SECI model. The key building blocks for knowledge management such as types of knowledge, specific knowledge management sub-processes, strategic options, tools to support knowledge management and different types of learning in project organisations are discussed. The business case for knowledge management and the elements for a knowledge management strategy such as resources and potential benefits are also briefly examined. The applications of practical tools developed in collaboration with leading design and construction firms for implementing KM strategy and benchmarking KM implementation efforts in project organisations such as CLEVER, IMPaKT and STEPS are described to show

how knowledge can be effectively managed in collaborative projects such as PPP/PFI projects.

6.2   Knowledge and Associated Concepts

In knowledge management literature, there is often a distinction made between data, information and knowledge. Data refer to raw numbers, discrete facts about events, whilst information is processed data that are analysed and structured within a particular context. Knowledge refers to the meaning of information in a specific context. It is the insight and experience that guide the thoughts, behaviours and decisions of people and the product of learning which is personal to an individual. Knowledge is about knowing what to do with information (actionable information or information with context) and can be categorised into internal and external knowledge, individual and group knowledge, explicit and tacit knowledge. 'Human activity is inconceivable without knowledge and the scope of knowing, and types of knowledge, are as wide and varied as all the varieties of human pursuits' (Quintas, 2005, p. 10). Marshall and Sapsed (2000) argued that 'organisations are depicted as storehouses of localised knowledge held by individuals and groups'. The key issue is therefore to identify localised knowledge and transform it into productive knowledge that resides within the organisation to create value (Stewart, 1997).

There are several classifications of knowledge types but perhaps the most widely used is the classification into explicit and tacit knowledge. Explicit knowledge is written down/stored in particular formats such as text, photographs, voice, video, and so on (e.g. specifications, engineering drawings, computer images of construction processes). It is reusable in a consistent and repeatable manner and therefore easy to transfer. Tacit knowledge, on the other hand, is stored inside people's heads and often 'we know much more than we can tell'. Tacit knowledge can be technical (such as the know-how of an expert) or cognitive – based on values, beliefs and perceptions – and it is therefore very difficult to transfer. According to Quintas (2005), 'Some forms of human knowledge can be communicated to others through language or symbols, such as the laws of thermodynamics or the names of the star constellations. Once codified, such knowledge is information or data that may be interpreted by others. Explicit or codified knowledge may be understood by people with complementary knowledge who can extract meaning from the "codes". Even this process of understanding or extracting meaning from information involves the use of tacit skills of interpretation, evaluation and generally making-sense of what is being conveyed. However, there is a dimension of knowledge which remains tacit and cannot be communicated in language or symbols.' He suggests that when there is a focus on codified knowledge, the scope is limited to no more than information management as the nature and scope of human knowledge are rather broader than that which can be encoded.

In addition to the tacit–explicit dimension, knowledge can also be viewed in other ways. For example, knowledge can exist within an organisation (internal knowledge) or outside an organisation (external knowledge). It can also be associated with a particular individual (individual knowledge) or groups of people within or outside an organisation (shared knowledge). It could also be classified based on the content of the knowledge – for example product knowledge, process knowledge, people knowledge, and so on. All of these can be used in combination with each other and with the tacit–explicit dimension, depending on the desired knowledge management activity or intervention.

6.3   Definitions and Perspectives of Knowledge Management

According to Quintas (2005), '[I]t is palpably obvious that without creating, accumulating, sharing and applying knowledge, no human civilization could have existed. Even though the phrase "knowledge management" only came into common usage in the West during the last five years of the 20th century, it is emphatically *not* the case that the management of organisational knowledge processes began in the mid 1990s.' He feels that the case studies of Honda, Matsushita and other firms in Nonaka and Takeuchi's influential and widely quoted book *The Knowledge Creating Company* (1995) were not examples of designated 'knowledge management' initiatives but rather descriptions of actual knowledge processes such as knowledge sharing, knowledge combination, and so on.

Knowledge management (KM) is a relatively new concept and there are many definitions, generally illustrating the variations in scope and content. KM is defined narrowly, in some instances, to emphasise the capture, access and reuse of information and knowledge using information technology. This definition implies that tacit knowledge can be converted to explicit knowledge using IT. However, it is generally accepted that tacit knowledge is more difficult to capture and manage. Examples of KM definitions are shown below:

- Knowledge management is the discipline of creating a thriving work and learning environment that fosters the continuous creation, aggregation, use and re-use of both organisational and personal knowledge in the pursuit of new business value. (Definition from Xerox Corporation. *Source*: Cross (1998))
- Knowledge management is 'any process or practice of creating, acquiring, capturing, sharing and using knowledge, wherever it resides to enhance learning and performance in organisations' (Scarbrough *et al.*, 1999).

KM is often defined from two main perspectives, namely, process perspective and outcome perspective. A process perspective definition considers KM as the process of controlling the creation, dissemination and utilisation of knowledge (Newman, 1991; Kazi *et al.*, 1999). Another process perspective definition considers KM as the '... identification, optimisation, and active management of intellectual assets, either in the form of explicit knowledge held in artefacts or as tacit knowledge possessed by individuals or communities to hold, share, and grow the tacit knowledge (Snowden, 1998).' The outcome perspective, on the other hand, focuses on the benefits that an organisation gets from managing its knowledge. An example is a definition that considers KM to be concerned with the way an organisation gains competitive advantage and builds an innovative and successful organisation (Kanter, 1999). Another example of an outcome perspective definition considers KM as the 'management of organisational knowledge for creating business value and generating competitive advantage' (Tiwana, 2000). A third example defines KM as 'the ability to create and retain greater value from core business competencies' (Klasson, 1999). A combined perspective defines KM by considering both its process and outcome. One example is that: 'Knowledge management enables the creation, communication, and application of knowledge of all kinds to achieve business goals' (Tiwana, 2000). Another definition states that KM is any process or practice of creating, acquiring, capturing, sharing and using knowledge, wherever it resides, to enhance learning and performance in organisations (Scarbrough *et al.*, 1999). Regardless of the different perspectives for defining KM, all definitions focus on the fact that knowledge is a valuable asset that needs to be managed and that managing this knowledge is important to improve organisational performance.

There are several other perspectives on KM. For example, there is the social/human resource perspective which focuses on KM as 'the distribution, access of human experiences and relevant information between related individuals and workgroups' (Excalibur Technologies, 1999). The IT/technology perspective focuses on the 'capture, access and reuse of knowledge using information technology' (O'Leary, 2001) and there is the intellectual capital/economic perspective which sees KM as the 'systematic and organised attempt to use knowledge within an organisation to transform its ability to store and use knowledge to improve performance' (KPMG, 1998).

Within the context of this chapter, KM can be simply defined as a systematic process of capturing, transferring and sharing knowledge to add competitive value (Drucker, 1993; Skyrme and Amidon, 1997; Hjertzen and Toll, 1999; Scarbrough and Swan, 1999) and to improve performance (Robinson *et al.*, 2001b). KM provides several benefits such as facilitating staff training, problem-solving and decision-making. It also enables the intellectual capital of an organisation (its skills, knowledge and processes) to be used effectively, creatively and consistently to improve business performance and customer satisfaction (TFPL Ltd, 1999). KM is therefore critical to an organisation's survival in competitive markets and it is becoming a strategic necessity for organisations willing to lead the market and even to those just wishing to keep their places in the market. The number of organisations that are implementing or planning to implement KM initiatives is increasing exponentially (Tiwana, 2000) because of the following reasons:

- Companies are becoming knowledge intensive rather than capital intensive.
- Unstable markets necessitate organised actions with regards to replacing old products and introducing new ones.
- KM allows companies to lead change.
- Only the knowledgeable organisations survive.

- Cross-industry amalgamation is already breeding complexity.
- Knowledge supports decision-making.
- Shared knowledge multiplies.
- Tacit knowledge can be lost easily.
- Competitors exist worldwide.

The focus of KM in collaborative project organisations such as PFI/PPP projects is to reflect various perspectives by using 'knowledge of best practice' whether tacit or explicit to improve project performance.

6.4   Theory of Knowledge Creation

Architects, engineers, surveyors, planners, facilities managers, clients and other legal, financial and technical specialists involved in PFI/PPP projects interact by using both codified and tacit knowledge during planning and de-sign development, construction, and operational and service delivery phases to deliver various outcomes. Explicit (codified) knowledge includes architec-tural design philosophy, engineering principles, design codes of practice and construction standards. Explicit knowledge is also captured or stored in var-ious documents to support PPP/PFI projects such as risk allocation matrix, value-for-money manuals and procedures for bidding in PFI/PPP projects, which are codified and easily communicated or shared with other members or people in the SPV/SPC. Tacit knowledge includes the experience of esti-mating the values of various risks in PFI/PPP projects, experience in tendering for PFI/PPP projects, practical design and work programming skills on PFI projects, all of which are acquired over a period of time.

According to Nonaka and Takeuchi's (1995) theory of knowledge cre-ation, there are four distinct modes of interaction that result in the cre-ation of knowledge (see the SECI Model in Figure 6.1). Construction project knowledge is created through the actions of individuals, project teams and construction organisations, and the interactions of these different types of knowledge (explicit and tacit) from conceptual design to the handover of the completed project.

Tacit-to-tacit interaction takes place through the process of socialisation. An architect giving a verbal account or an explanation of a design concept to a client during a meeting is an example of this type of interaction. Ap-prentice carpenters, bricklayers, plumbers, and so on, often work with their

	Tacit	Explicit
Tacit ➡	Socialisation (S)	Externalisation (E)
Explicit ➡	Internalisation (I)	Combination (C)

Figure 6.1 Knowledge creation theory.

masters to learn craftsmanship not through formal instruction but by sociali-sation which involves observation, imitation and practice. The long tradition of apprenticeship schemes in the construction industry is responsible for producing various craftsmen who rely on their tacit knowledge to solve con-struction problems. Such experiential knowledge is reinforced and developed through shared experience by continuous interaction and learning from each other. Similarly, young engineers, architects and surveyors supplement their academic training through mentoring. The mentors are often senior man-agement staff who can help individuals to learn, unlock their talents and to develop their knowledge in the organisation.

Internalisation takes place when knowledge is transformed from explicit to tacit by individuals. For example, an architect reading a textbook on design theory, or using a manual on design standards, could interpret these explicit documents to create an internal mental model of a unique design satisfying the clients' requirements and his or her taste and style. Externalisation is the reverse process where tacit knowledge is made explicit so that it can be shared. An architect engaged in a discussion with a contractor on-site, which is subsequently followed by a written instruction made available to specialist subcontractors, engineers and quantity surveyors is an example of an externalisation process. This process also takes place when an architect translates a design concept or mental model into sketches to explain to a client.

Explicit-to-explicit knowledge interaction takes place through a process called combination. Combination involves gathering, integrating, transfer-ring, diffusing and editing knowledge (Nonaka and Toyama, 2003). Individ-uals and project teams in construction create knowledge through integrat-ing and processing various project documents (e.g. design brief, sketches, project programme, engineering and production drawings, performance spec-ifications, conditions of contract, bills of quantities). Technologies such as e-mails, databases, CAD systems, document management systems and project extranets facilitate this mode of knowledge conversion.

Much of the training and experience of construction professionals is based on a balance between codified (explicit) knowledge and tacit knowledge. Case study interviews with structural design firms show that about 80% of knowledge used during concept design is tacit compared to about 20% of ex-plicit knowledge, whilst the reverse is true at the detailed design stage – 20% tacit and 80% explicit (Al-Ghassani, 2003). It is the dynamic interactions between tacit and explicit knowledge which facilitate decision-making in the implementation of construction projects. This is why construction project documents are understood and interpreted by those who have been through the same or similar type of training. For example, structural engineers could extract meanings from design codes or interpret construction drawings easily whilst accountants cannot.

Research conducted by den Hertog and Bilderbeek (1998) and Windrum *et al.* (1997) identified design, architecture, surveying and other construction services as knowledge-intensive service sectors. An important feature that dis-tinguishes knowledge-intensive sectors from manufacturing firms is the type of 'product' they supply and, following this, the role they play in regional and national innovation systems. Whereas manufactured products and processes contain a high degree of codified knowledge (they are 'commodification' of knowledge), knowledge-intensive sectors' are characterised by a high degree of tacit (intangible) knowledge. Specialised expert knowledge and problem-solving know-how are the real products of knowledge-intensive industries (Egbu and Robinson, 2005).

Professional knowledge (i.e. knowledge produced by consultants while interacting with their client's settings) is deeply embedded in a mutual socialisation process, where consultants and their clients design together their final output. This is often seen in the kinds of services provided by professional/consultancy firms of architects, quantity surveyors and engineers. For consultancy or professional firms, their main capital is intellectual assets, and most of their processes are geared towards the exploration, accumulation and exploitation of individual and firm expertise (Egbu and Robinson, 2005).

6.5   Types of Knowledge and Project Complexity

Construction projects are classified into three distinct types: standard construction, traditional construction and innovative construction (Bennett, 1991). Innovative projects are needed to satisfy the demands of clients with unusual needs, or where established answers are no longer appropriate as a result of market or technological changes (Bennett, 2000). Some PFI/PPP projects such as hospitals are very complex requiring a high degree of tacit knowledge from specialists and advisers from the private organisations participating as part of the SPV/SPC to provide solutions to meet the requirement of the public sector clients.

Process-based factors relate to the technical and management systems required for the delivery of PFI/PPP projects. Technical processes range from highly knowledge-intensive approaches relying heavily on tacit knowledge such as producing concept design using pencil and paper, and bidding during the planning and development phase to automated processes relying on intelligent and knowledge-based systems (explicit knowledge) codified in plant, machinery or robots for the construction phase. Standard construction projects are more effectively managed by programmed organisations relying heavily on routine and standard management procedures (codified knowledge) to manage the design and construction process. Innovative projects require a higher utilisation of tacit knowledge and flexible management structures to manage complex planning, design, construction and facilities management processes.

Standard design and construction projects require individuals with basic knowledge and skills. However, problem-solving or creative people are needed for complex PFI/PPP schemes requiring innovation such as major health care facilities that are difficult to plan, design and implement. Bennett (2000) argued that creative teams are needed for innovative projects as the variables to be considered are often ill-defined and the required technologies need to be developed. Creative or problem-solving teams are designed 'to bring knowledge to bear in solving *emergent problems*' (Dyer and Nobeoka,

2000). Team stability and the duration of traditional projects have a profound implication for knowledge creation and reuse. Egan (1998) noted that teams are disbanded at the end of every project and argued that 'the repeated selection of new teams inhibits learning, innovation and the development of skilled and experienced teams'. This view is supported by Bennett (2000) who argued that the best result comes from the same people working together project after project. In PFI/PPP projects, there is the opportunity for long-term collaboration, so new knowledge can be developed through the dynamic interaction of tacit and explicit knowledge to find solutions to client requirements.

6.6   KM Life Cycle

The KM life cycle consists of five distinct but interrelated sub-processes: discovery and capturing, organisation and storage, distribution and sharing, creation and leverage, and retirement and archiving (see Figure 6.2). The discovery and capturing stage is aimed at finding out where knowledge resides, whether in people' heads, processes or products. Examples include capturing tacit knowledge by bringing people together, discovering a database of products, experts or codified knowledge about processes. Knowledge organisation and storage deals with structuring, cataloguing and indexing knowledge so that retrieval can be done easily. An example includes the creation of database. Knowledge distribution and sharing is about getting the right knowledge to the right person or part of the organisation at the right time. It requires awareness of the relevant knowledge or best practice. Examples

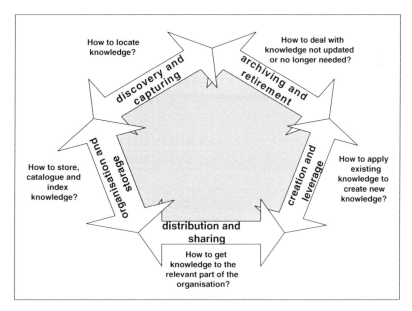

Figure 6.2   KM life cycle.

include using technology to distribute explicit knowledge or by connecting those who have tacit knowledge with those who need it.

The knowledge creation and leverage stage involves combining or applying knowledge in new ways to extend the overall knowledge of the business, and to exploit the new knowledge to improve business performance. Examples of knowledge creation include setting up project improvement and innovation processes to routinely monitor knowledge for new insight, whilst knowledge leverage includes licensing or selling knowledge through, for example, software products. Knowledge archiving and retirement stage deals with treatment of knowledge that has already been used but not updated or knowledge that has not been used or is no longer valid. This includes knowledge that is not of immediate use and relevance to the organisation but is placed in an archive to be retrieved as and when it becomes useful in the future. This stage is often ignored in the literature but it is becoming increasingly crucial in an era where information overload is a major problem.

6.7   KM Systems

A KM system is the technology platform and infrastructure that an organisation employs to support KM. It typically consists of a set of tools, made up of technologies (IT Tools) and techniques (non-IT tools). Both technologies and techniques are equally important to support different KM processes that are briefly described below.

6.7.1   KM technologies

KM technologies rely on an IT infrastructure. Examples of KM technologies for capturing knowledge are knowledge mapping tools, knowledge bases and case-based reasoning. Although there is a debate about the degree of importance of such technologies, many organisations consider these very important enablers that support the implementation of a KM strategy (Skyrme and Amidon, 1997; Kanter, 1999; Anumba *et al.*, 2000; Egbu, 2000; Storey and Barnet, 2000) as they consume one-third of the time, effort and money required for a KM system. The other two-thirds mainly relate to people and organisational culture (Davenport and Prusak, 1998; Tiwana, 2000).

KM technologies consist of a combination of hardware and software technologies. Hardware technologies and components are important for a KM system as they form the platform for software technologies to perform and are the medium for storage and transfer of knowledge. Some of the hardware requirements of a KM system include personal computers or workstations to facilitate access to knowledge, powerful servers to allow the organisation to be networked, open architecture to ensure interoperability in distributed environments, media-rich applications requiring Integrated Services Digital Network (ISDN) and fibre optics to provide high speed and use of the public networks (e.g. Internet) and private networks (e.g. intranet, extranet) to facilitate access to and sharing of knowledge (Lucca *et al.*, 2000). Software technologies play an important part in facilitating the implementation of KM. The number of software applications has increased considerably in the last few years. Solutions provided by software vendors take many forms and perform different tasks. The large number of vendors that provide KM solutions makes it extremely difficult to identify the most appropriate solutions. This has resulted in organisations adopting different models for establishing KM systems. Tsui (2002b) identifies five emerging models for deploying organisational KM systems where one or a combination may be adopted: customised off the shelf (COTS), in-house development, solution re-engineering, knowledge services, and knowledge marketplace.

KM software technologies have seen many improvements since the year 2000 due to many alliances and mergers and acquisitions between KM and Portal tool vendors (Tsui, 2002b). None of them, however, provides a complete solution to KM. These tools are better described within technology groups such as data and text mining and groupware. Ruikar *et al.* (2007) summarise the different KM software technologies in Table 6.1 (Haag and Keen, 1996; Haag *et al.*, 1998; Tsui, 2002a, 2002b).

6.7.2   KM techniques

KM techniques do not depend on IT although they provide support in some cases. Knowledge sharing, for example, is a sub-process of KM, which can take place through face-to-face meetings, recruitment, apprenticeships, mentoring and training. The importance of KM techniques comes from several factors. Firstly, KM techniques are affordable to most organisations as no sophisticated infrastructure is required. Some techniques, however, require more resources than others (e.g. training requires more resources than face-to-face interactions). Secondly, KM techniques are easy to implement and maintain due to their simple and straightforward nature. Thirdly, KM techniques focus on retaining and increasing the organisational tacit knowledge, a key asset to organisations.

KM techniques are not new; most organisations have been implementing these for a long time under the umbrella of management approaches such as organisational learning and learning organisations. Using these tools for the management of organisational knowledge requires their use to be enhanced so that benefits, in terms of knowledge gain/increase, can be fully realised. Examples of KM techniques include brainstorming, communities of practice (CoP), face-to-face interactions, post-project reviews, recruitment, mentoring, apprenticeship and training. Some of the KM techniques are more formal than others. On the one hand, there is face-to-face interaction which is useful for sharing the tacit knowledge owned by an organisation's employees. It is an informal and a powerful approach that helps in increasing the organisation's memory, developing trust and encouraging effective learning. Face-to-face interactions provide strong social ties and tacit-shared understandings that give rise to collective sense-making (Lang, 2001), which in turn leads to an emergent consensus as to what is valid knowledge and to the serendipitous creation of new knowledge and, therefore, new value. Training,

Table 6.1   KM software technologies and their uses (Ruikar *et al.*, 2007).

KM software technologies	Description and uses
Data and text mining	• Technology for extracting meaningful knowledge from masses of data or text • Enables identification of meaningful patterns and associations of data (words and phrases) from one or more databases or 'knowledge bases' • Enables identification of hidden relationships between data and hence creating new knowledge • Used in business intelligence, direct marketing and customer relationship management applications
Groupware	• Supports distributed and virtual project teams where team members are from multiple organisations and in geographically dispersed locations • Enables effective and efficient communication and sharing of information for geographically dispersed project teams
Intranet	• An internal organisational Internet that is guarded against outside access by special security tools called firewalls • Used for storing, sharing, accessing and locating company documents and information such as H&S standards, procedures, press releases, and so on
Extranet	• An intranet with limited access to outsiders, making it possible for them to collect and deliver certain knowledge on the intranet • Useful for making organisational knowledge available to geographically dispersed staff members
Knowledge base	• Repositories that store knowledge about a topic in a concise and organised manner • They present facts that can be found in a book, a collection of books, websites or even human knowledge. This is different from the knowledge bases of expert systems, which incorporate rules as part of the inference engine that searches the knowledge base to make decisions
Taxonomies and ontologies	• Taxonomy is a collection of terms (and the relationships between them) that are commonly used in an organisation. Examples of a relationship are 'hierarchical' (where one term is more general hence subsumes another term), 'functional' (where terms are indexed based on their functional capabilities) and 'networked' (where there are multiple links between the terms defined in the taxonomy) • Ontologies also define the terms and their relationships, but additionally, they support deep (refined) representation (for both descriptive and procedural knowledge) of each of the terms (concepts) as well as defined domain theory or theories that govern the permissible operations with the concepts in the ontology • Both can be used as corporate glossaries to hold detailed descriptions of key terms used in an organisation. They can also be used to constrain the search space of search engines and prune search results, identify and group people with common interests, and act as a content/knowledge map to improve the compilation and real-time navigation of web pages

on the other hand, usually follows a formal format and can be internal where seniors train juniors within the organisation, or external where employees attend courses managed by professional organisations or experts. The successful implementation of a training programme relies on careful planning and defined strategies (Ruikar *et al.*, 2007).

Other KM techniques such as brainstorming and CoP rely on groups for collective 'thinking' or problem-solving. Brainstorming, for example, requires a group of individuals to focus on a problem and intentionally propose as many deliberately unusual solutions as possible through pushing the ideas as far as possible. The participants discuss ideas and then build on these ideas. Only when the brainstorming session is over are the ideas evaluated (Tsui, 2002a, 2002b). Wenger (2000) defines a CoP as a social learning system, united by joint enterprise, mutually recognised norms and competence, with shared language, routines and stories. A community of practice is most often an informal grouping. It may be unrecognised (Scarbrough, 1996) or ignored or taken for granted in the organisation. Also, it may transcend organisational boundaries, including people in several organisations who hold experiences in common. Members of CoP may perform the same job or collaborate on a shared task (software developers) or work together on a product (engineers, marketers and manufacturing specialists). They are peers in the execution of 'real work'. What holds them together is a common sense of purpose and a real need to know what the other knows. Usually, there are many CoP within a single organisation and most people normally belong to more than one. CoP are sometimes referred to as knowledge communities, knowledge networks, learning communities, communities of interest or thematic groups (Ruikar *et al.*, 2007).

Post-project reviews facilitate the consolidation of learning and, to some extent, creating shared understanding on a project. They include ongoing (phase) reviews – both formal and informal (e.g. design reviews) – and post-project evaluation. One of the primary reasons for post-project evaluation cited by the Office of Government Commerce (OGC) is '*to transfer the knowledge and any lessons from one project to other projects*' (OGC, 2002). There are various guides on how to carry out project reviews (e.g. OGC, 2002 for public sector clients in the United Kingdom), and some organisations have well-established procedures. The effectiveness of such reviews undoubtedly depends not only on the way they are conducted, but also on the time allocated for reviews and the availability of the relevant staff (Kamara *et al.*, 2005).

6.8   Learning in Project Organisations

PFI/PPP project organisations operate in a dynamic and collaborative environment where there is a need to learn and share knowledge to facilitate continuous improvement. KM facilitates continuous improvement through project learning and innovation. From a project context, KM is a process of capturing, storing, sharing and applying the different types of knowledge, whether tacit or explicit, by making them easily accessible and usable so that time is saved, performance is improved and innovation is facilitated in the planning and design development, construction and operational phases of PFI/PPP projects. Project knowledge is defined here as the knowledge (including data and information) required to conceive, develop, realise and terminate a project (Kamara *et al.*, 2005). The project knowledge base is a function of the procedures put in place to transform knowledge. Knowledge can be captured from the diversity of people and specialists involved in PFI/PPP from public and private sector organisations, processes involved in PFI/PPP projects from planning and design development, construction through the operational and service delivery phases. However, the characteristics of PFI/PPP projects such as long-term cooperation and commitment, dynamic team membership from different disciplines in the SPV/SPC, and organisational boundaries raise a number of challenges for managing project knowledge and accelerating learning.

Project organisations have three distinct modes of learning: inter-project, intra-project learning and cross-sectoral or support learning. Inter-project learning takes place across projects by sharing lessons learned in previous projects to develop new knowledge for improving the performance of future projects. Documents relating to previous projects such as drawings, cost plans, bills of quantities, specification, work programme and project reports are often kept or archived for future references. In some cases, a summary of lessons learnt whether good or bad practices are also available following project closure. The scope for learning and sharing knowledge also depends on the type of project whether standard, traditional or innovative. The scope for sharing and reusing knowledge is greater in projects that rely on well-established standards at every stage of their design, manufacture and construction. It will also cost significantly less than non-standard projects as the benefit of reduced rework, reuse of drawings and reduction in uncertainty will improve project performance.

Intra-project learning takes place within a project by the creation and sharing of knowledge during the project life cycle. Intra-project learning provides an immediate and direct opportunity to influence an ongoing project as lessons learnt in earlier phases can be applied to subsequent phases for improvement. However, such benefits are not always fully realised as time is always a major constraint as a project progresses through different phases. Cross-sectoral or support learning takes place outside the project sector environment. There are a lot of good management practices and processes in manufacturing, aerospace and other sectors not used in the construction sector. There is therefore considerable scope for improvement in construction project organisations by looking at best practices in other industries. Egan (1998) argued that there is a need for a radical improvement in the construction industry, and suggested that the industry could learn from other industries to improve processes and product development. Research has been carried out on knowledge transfer across business sectors and it has been concluded that it is considerably less straightforward than commonly acknowledged (Fernie *et al.*, 2001).

At the planning and design development phase of PFI/PPP projects, the need for the project is established through various stages of developing the business case. Once the project is approved, the PFI/PPP project moves on from the advertisement, bidding, negotiating stages to the construction phase and finally the operational and service delivery phase. An important point to note about PFI/PPP projects is the complexity at the planning and design development phase as a result of different stakeholder interests and the dynamics between different project organisations involved. There are specific activities involved such as the preparation of the business case, the justification of the PFI/PPP approach by the public sector client and development of solutions from shortlisted bidders (made up of multidisciplinary teams) from the private sector based on the public sector client's output specification and other bid documents. For each activity in the different phases, there is an increased demand for knowledge, skills and learning, and the need to promote inter-organisational learning, requiring partners in the SPV/SPC to complement their knowledge and to develop an innovative solution for the public sector client.

Quintas (2005) argues that no firm has ever been independent in knowledge terms, but it is certainly the case today that all organisations are increasingly dependent on external sources of knowledge. This is particularly the case in PFI/PPP projects, which are by nature collaborative and multidisciplinary, and is exacerbated by the fact that not all members of the project team are involved in the project from start to finish. Thus, Quintas (2005) is of the view that organisations must develop absorptive capacity: the capability to access and assimilate new knowledge from external sources. Knowledge interdependence creates new management challenges resulting from the risks and difficulties of knowledge transactions across boundaries. Alliances, networks and collaborations (such as those exist in PFI/PPP projects) provide the means by which firms can reduce the risk, share costs and scarce resources, especially with regard to new or currently 'peripheral' technology areas (Quintas and Guy, 1995). However, the ability to share knowledge across functional and disciplinary boundaries presents particular KM challenges since different communities and disciplines may have little common ground for shared understandings (Quintas, 2005).

The successful transfer of knowledge between different projects is influenced by the way knowledge is captured (i.e. when and how) and repackaged (or codified) for reuse. Whatever process is set in place to achieve this should seek to the following:

- Facilitate the reuse of the collective learning on a project by individual firms and teams involved in its delivery.
- Provide knowledge that can be utilised at the operational and maintenance stages of the asset's life cycle.
- Involve members of the supply chain in a collaborative effort to capture learning in tandem with project implementation, irrespective of the contract type used to procure the project from the basis for both ongoing and post-project evaluation.

The issues of 'collective learning' and 'supply chain involvement' require concerted efforts to integrate the disparate stores of project knowledge for the mutual benefit of all project team members – this can be challenging (Kamara *et al.*, 2005).

In PPP/PFI projects where multidisciplinary parties and stakeholders are involved in financing, designing, constructing and maintaining a facility over a life cycle of 20–30 years, there is greater scope for effective knowledge transfer across project phases. Kamara *et al.* (2005) argued that it is important to note that the effectiveness of various contractual and/or organisational arrangements in facilitating cross-project knowledge transfer is dependent on whether the same people are used, or whether there is a strategy for sharing individual knowledge across the organisation. Improving cross-project knowledge transfer can be accelerated by implementing KM strategies at the level of the project organisation.

6.9   Developing a Business Case for KM

The emergence of the knowledge-based economy has been due to a number of factors such as changes in the global economic framework and the business model embracing knowledge as an organisational asset or the most important factor of production (intangible capital). It is now been recognised that the success of organisations increasingly depends on their ability to create new knowledge through effective KM strategies (Robinson *et al.*, 2005a). Many organisations are now realising some of the main business benefits of KM (Carrillo *et al.*, 2004):

- Reducing the loss of knowledge through turnover of staff/customers
- Decreasing cost of reinventing knowledge through dissemination of best practice
- Reducing the costs associated with repeating the same mistakes/rework
- Increasing the value of knowledge
- Responding to business opportunities more quickly

The strongest argument for developing a business case for KM is to demonstrate its business benefits so that the resources and support necessary for a successful implementation can be provided. For example, a leading engineering consulting organisation highlighted that feedback from their legal department shows that the single largest cause of loss of money within the firm was a failure to agree with the appropriate contract terms upfront (Sheehan, 2000). The knowledge manager explained that a KM system such as the collation of a legal intranet page pushed to the desktop at appropriate times in projects is an increasingly effective solution to this problem.

Specific KM initiatives will be required to facilitate the smooth running of PPP/PFI projects and to reduce the transaction costs of bidding and the whole life cost of PFI/PPP projects. Table 6.2 shows some cost savings that some leading organisations have achieved due to their KM programmes.

6.10   Development of a KM Strategy

The growing body of literature recommending how KM strategies could be developed (Storey and Barnet, 2000; Tiwana, 2000; Bollinger and Smith,

Table 6.2 Examples of cost savings from KM programmes (Robinson *et al.*, 2005b).

Texas Instruments saved itself the $500 million cost of building a new silicon wafer fabrication plant by disseminating best internal working practices to improve productivity in existing plants

Skandia AFS reduced the time taken to open an office in a new country from 7 years to 7 months by identifying a standard set of techniques and tools, which could be implemented in any new office

Dow Chemical has generated $125 million in new revenues from patents and expects to save in excess of $50 million in tax obligations and other costs over the next 10 years by understanding the value of its patent portfolio and actively managing these intellectual assets

Chevron Oil has made savings of $150 million per year in energy and fuel expenses by proactive knowledge sharing of its in-house skills in energy use management

---

2001) is opposed by the fact that developing methods and strategies for KM is a delicate task that is dependent on many factors. This explains why the recommendations made by these authors only describe KM strategies in very broad terms. Organisations' different cultures and different business goals make it impossible that one KM strategy, system or tool would suit every organisation. Furthermore, developing methods and strategies for implementing KM needs the integration of several issues such as people, culture and technology, which are usually unique to organisations. This means that proper planning is required in order to design robust KM strategies and systems.

There is a range of options that can be adopted to implement a KM strategy (Hansen *et al.*, 1999, Robinson *et al.*, 2001a). These include a personalisation strategy and a codification or computerisation strategy. In a personalisation strategy, the goal is to link people so that tacit knowledge is shared. The emphasis is on creating people sharing networks with IT providing a supporting role. In the codification/computerisation strategy, the goal is to connect people with reusable knowledge and the emphasis is therefore on capturing and reusing knowledge with IT playing a dominant role.

There are several elements to developing and implementing a KM strategy. These include the following:

- Deciding what key knowledge to share about processes, people and products/role of learning
- Deciding with whom to share (members of the SPV/SPC, internal and/or external organisations – e.g. suppliers, clients, individuals or groups of specialists)
- Deciding how to share (what KM tools – techniques or technologies – to use)
- Deciding which implementation issues to address – resources needed, reform (enabling conditions) and results monitoring systems
- Deciding how to evaluate the effectiveness and efficiency of KM strategy or initiatives

Appropriate consideration should also be given to the inputs or cost components of a KM initiative, otherwise it may not be implemented effectively. Components to consider include the following:

- *KM team component* represents the cost associated with the roles and skills required for knowledge transformation.
- *KM-enabled process component* represents the cost associated with core and supporting business processes affected or re-engineered.
- *KM infrastructure component* represents the costs associated with setting up human interactive systems to provide knowledge creation and sharing capability and the information and communication technology support to facilitate the knowledge transformation process.
- *Other costs component* such as management costs relating to change management programmes to address cultural issues and monitoring systems to support the implementation of the KM strategy.

There are different types of cost associated with the KM components such as staff (human) costs, organisational or (re)organisational costs, hardware and software costs. Cost allocations depend on the characteristics of the KM components. Costs could be direct or indirect, one-off/lump sum (e.g. purchase and initial installation cost of hardware and software, consultant's fee, etc.) or recurrent/periodic (e.g. hardware/software maintenance costs, staff costs, etc.) or occasional costs (e.g. hardware upgrades, support staff costs, etc.). As different KM tools are used for the implementation of KM initiatives, consideration should be given not only to their appropriateness in terms of functionality (i.e. ease of use, integration, focus and maturity) but also to cost. The cost component checklist is not prescriptive, as cost allocations will depend on the type of cost models used by individual organisations.

There are also different types of benefits to be expected, both tangible and intangible. Benefits from KM initiatives could also be direct or indirect (benefit contribution ratio to be determined where it is indirect). Operational benefits tend to be more direct or immediate benefits arising from the implementation of KM initiatives whereas strategic benefits are indirect, often realisable in the medium to long term. Evaluation of benefits could also be in the form of monetary units or utility values reflecting preference or degree of satisfaction/expectations in improvement.

The objective of evaluation is to identify the inputs (i.e. the nature of KM programmes) and their outputs (i.e. the consequences – both positive and negative) in terms of changes in performance measure or contribution to business benefits. Table 6.3 shows the various techniques used to assess the impact of KM initiatives.

There are also different types of evaluation – partial or full, *ex-ante* or *ex-post* evaluation. A *partial* evaluation involves a comparison of both input and output with respect to a single KM initiative (one alternative) or a comparison of at least two alternatives with respect to either their inputs or outputs. A *full* evaluation involves a comparison of both the inputs and outputs of at least two KM initiatives (two or more alternatives). An evaluation can also take place before implementation (*ex-ante* evaluation) or after implementation (*ex-post* evaluation) (Anumba *et al.*, 2002).

Table 6.3   Selection of evaluation techniques.

Evaluation technique	When to use it	Efficiency measure
Cost minimisation analysis: This involves a simple cost comparison of KM initiatives as it is assumed that the consequences (outputs) are identical or differences between the outputs are insignificant	When output of KM initiatives are identical in whatever unit of measurement is used	**Output** (constant)/**input** (monetary units)
Cost effectiveness analysis: This involves the comparison of KM initiatives where the consequences (output) are measured using the same natural or physical units	When output of KM initiatives are measured in the same natural or physical units, for example number of accidents prevented, reduction in absenteeism or waste, training man-hours, and so on	**Output** (many units)/**input** (monetary units)
Cost utility analysis: This involves a comparison of KM initiatives (inputs) which are measured in monetary units with the consequences (outputs) measured using utility or a preference scale	When a significant component of the output *cannot* be easily measured, quantified or expressed in monetary units Useful in making internal comparison between divisions within an organisation	**Output** (utility units)/**input** (monetary units)
Cost benefit analysis: This approach provides a comparison of the value of input resources used up by the KM initiative compared to the value of the output resources the KM initiative might save or create. Consequences of KM initiatives are measured in monetary terms so as to make them commensurate with the costs.	When a significant component of the output *can* be easily measured, quantified or expressed in monetary units Useful in determining return on investment (ROI), Internal Rate of Return (IRR), Net Present Value (NPV) or Payback Period of KM investments	**Output** (monetary units)/**Input** (monetary units)

All of the above considerations need to be an integral part of the development of a KM strategy. There are a number of tools that can help with this and some of them are discussed in the next section as part of the KM toolkit for PFI/PPP projects.

6.11   KM Toolkit for PFI/PPP Projects

There are a variety of tools to support KM implementation. However, most of the current KM systems tend to focus on requirement analysis, design development and implementation. Gallupe (2001) is of the view that KM

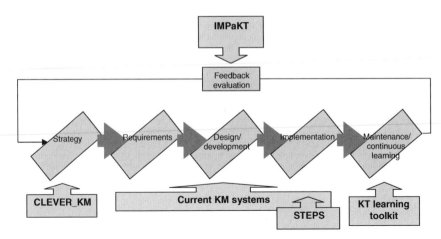

Figure 6.3   KM solution road map.

tools should not be considered synonymous with information management tools as they should be 'capable of handling the richness, the content, and the context of the information and not just the information itself'. The term 'KM tools' is sometimes used narrowly to mean information technology (IT) tools. There are many competing products in the marketplace, but there are problems of overlap between the functions of various tools and the high cost of acquiring and using them. There are also a number of gaps that the existing tools do not address – some of these are vital in PFI/PPP projects.

Various tools have been developed by the Construction Informatics Research Group at Loughborough University to address the aforementioned gaps. These include CLEVER_KM™ IMPaKT, STEPS and the Knowledge Transfer (KT) Learning Toolkit, which were developed to facilitate the development of KM strategies, evaluation and benchmarking of KM initiatives, assessment of the maturity of KM in organisations and to enhance knowledge transfer capacity in joint venture projects (such as PFI/PPP). Figure 6.3 illustrates how these tools – CLEVER, IMPaKT, STEPS and the KT Learning Toolkit – fit into the KM solution road map. These are briefly described below and it is evident that they can facilitate the development of a total KM and capacity building solution.

> **CLEVER_KM:** The CLEVER_KM tool provides a structured approach to KM problem definition and strategy formulation based on specific steps shown in Figure 6.4. The CLEVER tool consists of several stages that take an organisation from an initial definition of a knowledge problem, an identification of where they wish to get to, specification of the critical migration paths required, through to the provision of appropriate KM processes to aid in the resolution of the organisation's KM problem (Anumba *et al.*, 2005). The key features of each stage in the framework are briefly described below:
>
> > **Define KM problem:** This involves a description of the perceived KM problem and identification of the underlying business drivers. Use

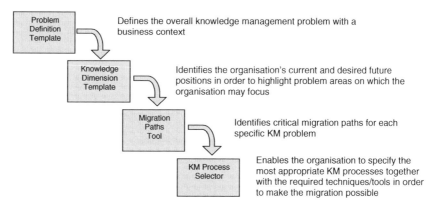

Figure 6.4   Steps in the CLEVER_KM tool.

is made of a Problem Definition Template (PDT) to characterise the knowledge under consideration and to establish the potential sources and means of acquiring it.

Identify 'to be' solution: This stage is used to confirm the characteristics of the current (as-is) position and to identify the desired (to-be) position for each problem area with regard to an organisation's strategy and policy. Use is made of a Knowledge Dimensions Guide, which is a sliding scale with predefined 'states' on which the organisation can specify their current and desired positions.

Identify critical migration paths: This stage focuses on how the organisation wishes to proceed from its current position to the desired position. Use is made of a set of matrices (the Migration Paths Tool) that define the implications of various migration options.

Select appropriate KM processes: At this stage, the organisation is guided through the selection of appropriate KM processes to enable them to move along each migration path. Thus, depending on the chosen migration path, the most relevant KM processes are identified such that the organisation can develop action plans.

The CLEVER_KM system has been commercialised and is useful for KM strategy formulation at both organisation and project organisation level.

IMPaKT: The IMPaKT framework (shown in Figure 6.5) enables organisations to evaluate the business impact of their KM initiatives (Carrillo *et al.*, 2003; Robinson *et al.*, 2004). The framework facilitates (1) an understanding of the strategic context of business problems and their KM implications, (2) the planning and alignment of KM strategy to address business problems or objectives so as to ensure that they are coherent and consistent with the overall strategic objectives of an organisation and (3) an evaluation of the impact of KM on business performance in terms of effectiveness and efficiency by providing a set of complementary measures for assessing the impact of KM initiatives on business performance. Evaluation with industrial practitioners showed that the framework could significantly facilitate the implementation of a KM

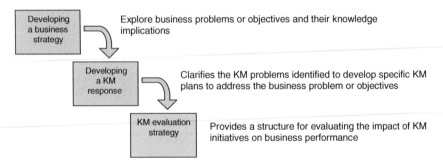

Figure 6.5   The IMPaKT framework.

strategy in construction project organisations. The framework enables a useful thought process, is well focused and easy to use, and is considered to be an innovative KM tool that incorporates issues not adequately addressed in other frameworks.

STEPS: The STEPS maturity road map provides a structured approach to determine the steps and action plan required and to benchmark KM implementation efforts (Robinson *et al.*, 2006). The five steps in the maturity road map (start-up, take-off, expand, progress and sustain) reflect varying levels of KM maturity. Each level of maturity is characterised or associated with certain attributes and attribute dimensions. Key aspects of the road map are shown in Figure 6.6, reflecting the different emphases at various stages, as detailed below.

The KT Learning Toolkit: The KT learning and capacity building toolkit (Carrillo *et al.*, 2006) was specifically developed for PFI/PPP projects to encourage organisations to transfer knowledge on all aspects of the

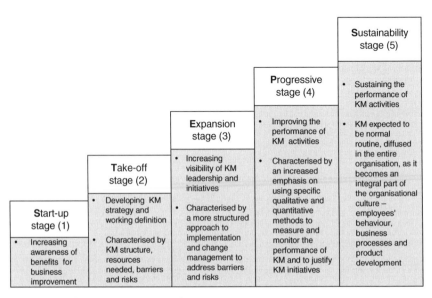

Figure 6.6   The STEPS maturity road map.

PFI/PPP process. It involves a three-stage process of (1) assessing business opportunities to improve participation in PFI/PPP projects, (2) building a knowledge map and transfer capability and (3) creating an action plan for learning and capacity building. The details of how the knowledge transfer framework operates to enhance learning and capacity building for PFI/PPP projects are provided in Chapter 9 of this book.

6.12   Concluding Remarks

KM is very important for the success and continuous improvement in PFI/PPP projects. It offers numerous benefits that can help to increase the returns for firms in the SPV/SPC that deliver PFI projects. However, the long-term and multi-organisation nature of PFI/PPP projects present additional challenges to the management of knowledge in these project environments. This chapter has sought to provide some guidance on these issues. It started with a general discussion of key KM principles, theories and concepts, and then moved on to address aspects of KM that have a particular resonance in the PFI/PPP project context – tools and techniques, learning in project organisations, business case for KM, KM strategy development, and so on – and presents a KM toolkit that will enable organisations involved in PFI/PPP projects to better manage their corporate and collective knowledge. The next chapter builds on this chapter by presenting case studies of knowledge transfer on PFI/PPP projects.

References

Al-Ghassani, A.M. (2003) *Improving the Structural Design Process: A Knowledge Management Approach*. PhD Thesis. Loughborough University, Leicestershire.

Anumba, C.J., Bloomfield, D., Faraj, I., and Jarvis, P. (2000) *Managing and Exploiting Your Knowledge Assets: Knowledge Based Decision Support Techniques for the Construction Industry, BR382*. Construction Research Communications Ltd, UK.

Anumba, C.J., Carrillo, P.M., Robinson, H.S., and Al-Ghassani, A.M. (2002) *A Performance-Based Approach to Knowledge Management*. In: Rezgui, Y., Ingirige, B., and Aouad, G. (eds). Proceedings of the eSMART Conference. University of Salford, 19–21 November 2002, Part A, pp. 134–145.

Anumba, C.J., Kamara, J.M., and Carrillo, P.M. (2005) In: Anumba, C.J., Egbu, C., and Carrillo, P.M. (eds). Knowledge management strategy development: a CLEVER approach. *Knowledge Management in Construction*. Blackwell, Oxford, pp. 151–169.

Bennett, J. (1991) *International Construction Project Management: General Theory and Practice*. Butterworth-Heinemann, Oxford.

Bennett, J. (2000) *Construction – The Third Way, Managing Cooperation and Competition in Construction*. Butterworth-Heinemann, Oxford.

Bollinger, A.S., and Smith, R.D. (2001) Managing organisational knowledge as a strategic asset. *Journal of Knowledge Management* 5(1), 8–18.

Carrillo, P.M., Robinson, H.S., Al-Ghassani, A.M., and Anumba, C. (2004) Knowledge management in construction: drivers, resources and barriers. *Project Management Journal* 35(Part 1), 46–56.

Carrillo, P.M., Robinson, H.S., Anumba, C.J., and Al-Ghassani, A.M. (2003) IM-PaKT: a framework for linking knowledge management to business performance. *Electronic Journal of Knowledge Management (EJKM.)* 1(1), 1–12.

Carrillo, P.M., Robinson, H.S., Anumba, C.J., and Bouchlaghem, N.M. (2006) Knowledge transfer framework: the PFI context. *Construction Management and Economics* 24(10), 1045–1056.

Cross, R. (1998) Managing for knowledge: managing for growth. *Knowledge Management* 1(3), 9–13.

Davenport, T.H., and Prusak, L. (1998) *Working Knowledge: How Organisations Manage What They Know*. Harvard Business School Press, Boston.

den Hertog, P., and Bilderbeek, R. (1998) *Innovation in and through Knowledge Intensive Business Services in The Netherlands*. TNO-report STB/98/03, TNO/STB 1997.

Drucker, P. (1993) *Post-Capital Society*. Butterworth-Heinemann, Oxford.

Dyer, J.H., and Nobeoka, K. (2000) Creating and managing a high performance knowledge sharing network: the Toyota case. *Strategic Management Journal* 21, 345–367.

Egan, J. (1998) *Rethinking Construction: Report of the Construction Task Force on the Scope for Improving the Quality and Efficiency of the UK Construction Industry*. Department of the Environment, Transport and the Regions, London.

Egbu, C.O. (2000) *Knowledge Management in Construction SME's: Coping with the Issues of structure, Culture, Commitment and Motivation*. Proceedings of the 16th Annual Conference of ARCOM, 6–8 September, Vol. 1. Glasgow Caledonian University, UK, pp. 83–92.

Egbu, C.O., and Robinson, H.S. (2005) Construction as a knowledge based industry. In: Anumba, C.J., Egbu, C., and Carrillo, P.M. (eds). *Knowledge Management in Construction*. Blackwell, Oxford, pp. 10–30.

Excalibur Technologies (1999) *Knowledge Retrieval – The Critical Enabler of Knowledge Management*. An Excalibur Technologies White Paper, Excalibur Technologies International Ltd. Berkshire.

Fernie, S., Weller, S., Green, S.D., Newcombe, R., and Williams, M. (2001) *Learning Across Business Sectors: Context, Embeddedness and Conceptual Chasms*. Proceedings of the ARCOM 2001 Conference, Vol. 1. University of Salford, UK, 5–7 September, pp. 557–565.

Gallupe, R.B. (2001) Knowledge management systems: surveying the landscape. *International Journal of Management Reviews* 3(1), 61–67.

Haag, S., Cummings, M., and Dawkins, J. (1998) *Management Information Systems for the Information Age*. McGraw-Hill, Boston, MA.

Haag, S., and Keen, P. (1996) *Information Technology: Towards Advantage Today*. McGraw-Hill, New York.

Hansen, M.T., Nohria, N., and Tierney, T. (1999) *What's Your Strategy for Managing Knowledge*. Harvard Business Review, March–April, pp. 106–117.

Hjertzen, E., and Toll, J. (1999) *Measuring Knowledge Management at Cap Gemini AB*. MSc Dissertation. Linkoping University, Sweden.

Kamara, J.M., Anumba, C.J., and Carrillo, P.M. (2005) Cross-project knowledge management. In: Anumba, C.J., Egbu, C., and Carrillo, P.M. (eds). *Knowledge Management in Construction*. Blackwell, Oxford, pp. 103–120.

Kanter, J. (1999) Knowledge management, practically speaking. *Information Systems Management* 16(4), 7–15.

Kazi, A.S., Hannus, M., and Charoenngam, C. (1999) *An Exploration of Knowledge Management for construction.* In: Hannus, M. et al. (eds). Proceedings of the 2nd International Conference on CE in Construction (CIB Publication 236). Espoo, Finland, 25–27 August, pp. 247–256.

Klasson, K. (1999) Managing knowledge for advantage: content and collaboration technologies. *The Cambridge Information Network Journal* 1(1), 33–41. [Cited in: Tiwana, A. (2000) *The Knowledge Management Toolkit.* Prentice Hall Inc., New Jersey.]

KPMG Management Consulting (1998) *Knowledge Management Research Report.* KPMG Management Consulting, Chicago, IL.

Lang, J.C. (2001) Managerial concerns in knowledge management. *Journal of Knowledge Management* 5(1), 43–57.

Lucca, J., Sharda, R., and Weiser, M. (2000) Coordinating Technologies for Knowledge Management in Virtual Organisations. *Proceedings of the Academia-Industry Working Conference on Research Challenges CAIWORC'00, 27–29 April.* Buffalo, New York.

Marshall, N., and Sapsed, J. (2000) The Limits of Disembodied Knowledge: Challenges of InterProject Learning in the Production of Complex Products and Systems. *Knowledge Management: Concepts and Controversies Conference, University of Warwick, 10–11 February.*

Newman, B.D. (1991) *An Open Discussion of Knowledge Management.* The Knowledge Management Forum. Available online at www.3-cities.com/~bonewman/what_is.htm (Accessed 5 October 2001).

Nonaka, I., and Takeuchi, H. (1995) *The Knowledge Creating Company.* Oxford University Press, New York.

Nonaka, I., and Toyama, R. (2003) The knowledge creating theory revisited: knowledge creation as a synthesising process. *Knowledge Management Research and Practice* 1, 2–10.

O'Leary, D.E. (2001) How Knowledge Reuse Informs Effective System Design and Implementation. *IEEE Intelligent Systems* 16(1), 44–49.

Office of Government Commerce (OGC) (2002) *Project Evaluation and Feedback.* Available online at www.ogc.gov.uk/sdtoolkit/reference/achieving/acieving.html#lib9 (Accessed June 2003).

Quintas, P. (2005) The nature and dimensions of knowledge management. In: Anumba, C.J., Egbu, C., and Carrillo, P.M. (eds). *Knowledge Management in Construction.* Blackwell, Oxford, pp. 10–30.

Quintas, P., and Guy, K. (1995) Collaborative, pre-competitive R&D and the firm. *Research Policy* 24, 325–348.

Robinson, H.S., Anumba, C.J., and Carrillo, P.M. (2006) STEPS: a knowledge management maturity roadmap for corporate sustainability. *Business Process Management Journal (Special Issue on Managing Business Processes for Corporate Sustainability)* 12(6), 793–808.

Robinson, H.S., Carrillo, P.M., Anumba, C.J., and Al-Ghassani, A.M. (2001a) Knowledge Management: Towards an Integrated Strategy for Construction Project Organisations. *Proceedings of the Fourth European Project Management (PMI) Conference, London, 6–7 June, Published on CD.*

Robinson, H.S., Carrillo, P.M., Anumba, C.J., and Al-Ghassani, A.M. (2001b) Linking Knowledge Management to Business Performance in Construction Organisations. *Proceedings of ARCOM 2001 Conference, Salford, United Kingdom, 5–7 September,* pp. 577–586.

Robinson, H.S., Carrillo, P.M., Anumba, C.J., and Al-Ghassani, A.M. (2004) Developing a business case for knowledge management: the IMPaKT approach. *Construction Management and Economics* 22(7), 733–743.

Robinson, H.S., Carrillo, P.M., Anumba, C.J., and Al-Ghassani, A.M. (2005a) Knowledge management practices in construction organisations. *Engineering, Construction and Architectural Management* 12(5), 431–445.

Robinson, H.S., Carrillo, P.M., Anumba, C.J., and Al-Ghassani, A.M. (2005b) Performance measurement in knowledge management. In: Anumba, C.J., Egbu, C., and Carrillo, P.M. (eds). *Knowledge Management in Construction*. Blackwell, Oxford, pp. 132–150.

Ruikar, K., Anumba,C.J., and Egbu, C. (2007) Integrated use of technologies and techniques for construction knowledge management. *Knowledge Management Research and Practice* 5, 297–311.

Scarbrough, H. (1996) *Business Process Re-design: The Knowledge Dimension*. ESRC Business Processes Resource Centre, University of Warwick, UK.

Scarbrough, H., and Swan, J. (1999) *Case Studies in Knowledge Management*. Institute of Personal Development (IPD), London.

Scarbrough, H., Swan, J., and Preston, J. (1999) *Issues in People Management: Knowledge Management: A Literature Review*. Institute of Personnel and Development, The Cromwell Press, Wiltshire.

Sheehan, T. (2000) Building on knowledge practices at Arup. *Knowledge Management Review* 3(5), 12–15.

Skyrme, D.J., and Amidon, D.A. (1997) *A Report on Creating the Knowledge-Based Business*. Business Intelligent Limited, London.

Snowden, D. (1998) In: Rock, S. (ed.) *A Framework for Creating a Sustainable Programme, Knowledge Management: A Real Business Guide*. Caspian Publishing, London, pp. 6–18.

Stewart, T.A. (1997) *Intellectual Capital: The New Wealth of Nations*. Doubleday, New York.

Storey, J., and Barnet, E. (2000) Knowledge management initiatives: learning from failure. *Journal of Knowledge Management* 4(2), 145–156.

TFPL Ltd (1999) *A Report on Skills for Knowledge Management – Building a Knowledge Economy*, 1st edn. London.

Tiwana, A. (2000) *The Knowledge Management Toolkit: Practical Techniques for Building a Knowledge Management System*. Prentice-Hall, New Jersey.

Tsui, E. (2002a) *Technologies for Personal and Peer to Peer (P2P) Knowledge Management, CSC Leading Edge Forum Technology Grant Report*. Available online at http://www2.csc.com/lef/programs/grants/finalpapers/tsui_final_P2PKM.pdf (Accessed 6 June 2002).

Tsui, E. (2002b) In: Holsapple, C.W. (ed.) *Tracking the Role and Evolution of Commercial Knowledge Management Software, in Handbook on Knowledge Management*. Springer-Verlag, Berlin/Heidelberg.

Wenger, E. (2000) Communities of practice and social learning systems. *Organisation* 7(2), 225–246.

Windrum, P., Flanagan, K., and Tomlinson, M. (1997) *Recent Patterns of Services Innovation in the UK*. Report for TSER project 'SI4S'. PREST, Manchester.

7 Case Studies on Knowledge Transfer

One of the central pillars of knowledge management concerns learning from others about their experiences and how they have overcome their problems. The previous chapter examined knowledge management theory, the role of knowledge management and learning in project-based organisations, the business case for developing a knowledge management strategy and the need for benchmarking knowledge management implementation efforts.

This chapter examines some of the key issues relating to the implementation of PPP/PFI projects and application of knowledge management strategy from the perspective of various PFI/PPP stakeholders. A case study approach is adopted to provide key learning points from the experience of (1) a public sector client, (2) a special purpose vehicle company, (3) a consultant and adviser, and (4) a design and build contractor and facilities management provider. For each case, the company's strategy towards PFI/PPP, knowledge issues arising during key PFI stages considered critical by industrial collaborators and their knowledge transfer strategy are examined. The three key stages identified by industrial collaborators as critical are the outline business case (OBC) stage, preferred bidder (PB) stage and facilities management (FM) stage. The chapter examines the specific issues arising from the case studies such as the potential scope for learning, improvement and organisational readiness and the need for knowledge transfer within these organisations.

7.2 Perspective of Public Sector Client

Case Study 7.1: A Public Sector Client

Case Study 7.1 is a PFI client organisation employing over 20 000 people with an annual turnover of £600 million. They are promoters who are interested in improving their PFI processes as they have implemented PFI schemes in the past and are embarking on another project. Case Study 7.1's main motivation for involvement in PFI is to access significant funding as this is the only way to provide large projects that implement the government's programme for refurbishing schools.

PFI/PPP strategy

Case Study 7.1 does not have a structured PFI/PPP strategy. Their approach is opportunistic from individual departments and there was no evidence of a strategy being developed at the time. Their main PFI/PPP activities not only focus on consolidating in schools/education sector but they are also looking at other opportunities such as waste PFI/PPP schemes. The local authority is driven by the need to continue raising standard of education and having better buildings is part of that. Individual departments are supported by elected members and no one has been appointed to look at strategy at a corporate level. The first PFI project had a dedicated Project Manager with two full-time staff and widespread support from people in various departments such as estates, legal, planning and finance. Their second PFI scheme is likely to have five staff with external advisers providing legal, financial and planning advice. There is also a separate budget identified for consultants and other costs to develop the project. The organisation has access to main IT infrastructure provided for the entire county but there will be additional IT facilities (Electronic Document Management System) for the second PFI project. This will not be used for drawings but all the other documentation involved such as planning applications, minutes of meetings, project documents such as Invitation to Negotiate documents, and so on.

Most of the framework for schools PFI/PPP schemes, now called Building Schools for the Future (BSF), comes from the then Department for Education and Skills (DfES) (now the Department for Children, Schools and Families (DCSF)) who set the parameters, that is central government objectives for the bids, criteria for judging the bids, and any scheme has to have an approval before from the Government Department (DCSF) being put out to tender. Information and support are also provided from the Public-Private Partnership Programme (4Ps), a body funded by the local government association that runs network activities that offer opportunities for the exchange and sharing of experience in PFI/PPP projects and also supports individual local education authorities (LEA) in need. Networking occurs through the Education Building and Development Officers group which is a standing conference for discussing a variety of topics relating to school building procurement, not just PFI/PPP projects.

The key enablers are getting elected members (councillors') support which is crucial as well as getting schools and governors to sign up to it. The main barriers are ongoing affordability problems mainly due to life cycle costs – the biggest single contributor to the affordability gap. There were no specific performance measures set in the first scheme. Performance measures to assess impact of schemes include improvement in GCSE pass rates, 'A' level results, pupil attendance, staff retention, absenteeism and illnesses. BSF is particularly targeted at areas of disadvantage to improve exam results and school attendance. The local authority argued that there is a real excitement about the new facilities that have been built under the first PFI/PPP scheme. For example, 'the science labs at the old schools were dreadful and have been replaced with high quality facilities'.

Knowledge issues during PFI stages

Case Study 7.1 felt that there is a role for knowledge transfer, particularly in trying to help heads of schools and governors to understand what to expect during the construction and service delivery phases and their rights and responsibilities under PFI contracts. Case Study 7.1 assesses its OBC stage as highly significant from a client's point of view because if they do not get this stage right the scheme will not be approved. A key knowledge required at OBC is identifying expert advice or experience. Network groups are useful in sharing knowledge and 4Ps facilitate these networks. Understanding the Gateway Review process/peer-review scheme is also important. The OBC, which includes the financial modelling and education case, has to go through the DfES, now called DCSF, then the project review group which meets every 6 weeks to assess OBC. The key activities from the client's perspective at the OBC stage include the following:

- Doing more work on the costing/ scope (what will go into the scheme) and school budgets.

- Getting financial advisers to do the modelling (capital cost, revenue, public sector comparator – PSC) and life cycle costs – affordability. There are various versions of financial models – do benchmarking and look at cleaning costs based on areas.
- Proving that pupil numbers are sustainable. Using optimism bias technique to adjust figures for the fact that people tend to underestimate cost and overestimate delivery (see the Treasury Green Book for details).
- Demonstrate value for money – the value of risk transferred to the private sector is absolutely critical.
- Better estimate of affordability gap based on PFI credits (for capital cost), school budgets (for ongoing FM requirements) and any capital receipts (e.g. sale of land). This is the equivalent of the mortgage analogy – more capital reduces your monthly outgoings.
- Demonstrate that elected members understand the affordability gap and are prepared to underwrite it.
- Look at any project management arrangements the authority has in place. There is a need for a Project Manager with delegated authority.

Table 7.1 provides a summary of the key problem areas, a description of the problem areas, staff to be targeted for learning and scope for learning for each of the three stages.

Table 7.1   Case Study 7.1 knowledge issues at outline business case stage.

Key problem areas	Description	Staff to be targeted	Scope for learning
Unrealistic budget	Staff now understand that the budget available is inadequate	Education department (finance staff) Building services and estates External financial advisers	Moderate: People need to interrogate numbers better
Poor historical data	Things have improved in schools. Every authority now has an asset management plan. Planning and access (property/land issues) need to be done earlier	Education department (finance staff) Building services and estates External financial advisers	Moderate
Too many assumptions	This is a problem but there is now a body of knowledge/benchmarking. It is easier now to get more accurate data	Financial adviser	Limited: Assumptions now have a sounder basis so it is not a big problem
Confidence in the financial data	Experienced advisers are crucial as most authorities do not have the capacity to do financial modelling	Client (internal team) Advisers	Limited

The PB stage is also considered as vital and is also rated highly significant as if it goes wrong, the PFI/PPP deal will not be signed. There is the need for specialist knowledge at the PB stage – lawyers, architects and HR people for support so that the scheme can be finalised satisfactorily to go ahead. The public sector client is aware that any signed deal will last for 25 years, so it is important to select the right partner. The key activity at the PB stage from the client's perspective is the evaluation of bids and includes splitting it down into particular areas:

- Response to project agreement
- Financial model/costing and affordability
- Design (does it deliver what you want, DCSF statutory requirements)
- Service proposals

Table 7.2 provides a summary of the key problem areas, a description of the problem areas, staff to be targeted for learning and scope for learning for each of the three stages.

Table 7.2   Case Study 7.1 knowledge issues at PB stage.

Key problem areas	Description	Staff to be targeted	Scope for learning
Redesign scheme due to affordability	Design would have been developed over time so ought not be a problem at this stage	Local authority (internal architects) Schools	Moderate
Contractual issues	It is a challenge to get an unambiguous commitment (standard contract in education)	Legal advisers	Moderate
Transfer of personnel	Trade unions unsure of who they are going to work for	HR department Unions	Moderate – used to be much more contentious
Town planning issues	Ensure that outline planning is in place and confident about detailed planning. Detailed planning is the PB's responsibility	Client Advisers SPV/designers	Moderate
Land assembly	Is there any land to buy to get access to the site or to provide an adequate site area to meet guidelines? Are there land disposals as part of the scheme?	Client Advisers (financial, legal and technical)	Moderate

The key role of the local authority is to support schools and get them to understand what they are entitled to in the PFI/PPP contract, what processes they need to go through if they want to change any aspect of the service. The main contract is between the local authority and special purpose vehicle, but there is an indirect relationship with the FM provider as the FM provider will have the most day-to-day contact with the schools. There is a particular need for knowledge at FM stage to enable school governors to be able to interpret drawings which may require training so that they can understand what is in a contract, and advice given from DCSF and lawyers.

Table 7.3 provides a summary of the key problem areas, a description of the problem areas, staff to be targeted for learning and scope for learning for each of the three stages.

Knowledge transfer strategy

The organisation felt that there is a role for knowledge transfer, particularly in trying to help heads of schools and governors to understand what to expect during the construction and service delivery

Table 7.3 Case Study 7.1 knowledge issues at FM stage.

Key problem areas	Description	Staff to be targeted	Scope for learning
Defects leading to problems with FM	Evaluate the involvement of the FM provider earlier on	Facility Managers Design and build contractors	Limited
Design standards and visualisation	Presentation by architects/design team will help with standards and visualisation	Local authority Architects Clients School governors	Limited
Managing user expectations	People tend to focus too much on design and not enough on what happens when the users start using the facility	Local authority User groups	Significant
Change of building legislation	Results in a change in insurance requirements	Government Insurance company	Moderate

phases and their rights and responsibilities under PFI contracts. There is no knowledge transfer strategy available internally, so this is where the role of 4Ps is crucial. There are always initiatives to improve the way local authorities work such as best value, comprehensive performance assessment, but no specific improvement strategy for PFI. There is also an Ofsted framework to assess educational authorities.

Learning mechanisms include networking groups such as 4Ps, conferences and talking to advisers. There is nothing formal to measure improvement in learning or the mechanisms used. The organisation felt that there is a need for a knowledge transfer strategy to make lessons learned available. This is crucial between authorities. 4Ps provides support for that within authority.

7.3   Perspective of Special Purpose Vehicle (SPV)

Case Study 7.2: A Special Purpose Vehicle

Case Study 7.2 has had a number of strategic disposals in recent years and now concentrates solely on PPP's and in particular PFI schemes. Case Study 7.2 already had built up a number of Design, Build, Finance and Operate (DBFO) schemes since the 1970s which has helped it to become a market leader in PFI.

Case Study 7.2 is a market leader in several sectors of the PFI accommodation market, including defence, health care, emergency services and education. The company has won several awards for its significant growth, and for becoming the leading sponsor of PFI/ PPP projects.

Case Study 7.2's core business is PPP with the focus on bidding, investing and managing PFI schemes as well as building up a significant operations arm through the provision of FM services. The company also operates PFI schemes in the United Kingdom, Australia and Europe. The key factors in determining its continuous involvement are (1) shareholders returns, (2) competitive edge in bid management and (3) FM operations as their resources are geared to PPP.

PFI/PPP strategy

Case Study 7.2's strategy is to have large, dedicated teams for each of the various sectors and manages its portfolio with constraints in mind. It is also selective about bidding for new projects based on availability of partners and teams, and government priorities (e.g. health, hospitals, LIFT). The overall strategy is to build a profitable portfolio which in some cases involves buying assets from others. The company has plans to continue their involvement in health and education as priority areas. Criminal justice – courts and police (not prisons as they do not have a competitive advantage there), defence – as they have two joint venture companies and regeneration (social housing and community facilities) – are other areas they are involved in. They also intend to continue expansion into roads and railways which are other parts of the businesses. There are contingency plans for the future which includes expanding into Portugal, Spain, Germany and Scandinavia.

Case Study 7.2 employs about 1200 people; the vast majority of which are full-time. There is an intranet, an Electronic Document Management System (EDMS) and an overall budget for each sector. Employees are forced to save everything into the EDMS to encourage sharing. There are no individual files for saving information but a central filing system/repository database. Almost all the schemes Case Study 7.2 is involved in are PFI with the exception of a few property based PPP's and some rail projects which involve securing franchising.

The company measures the benefits of participating in PFI/PPP schemes by the expected commercial returns as shareholders continue to rely on this. PFI/PPP projects managed by each division within the group are assessed according to the objectives and aim of each scheme. Performance measures include return on equity, avoidance of penalties, minimisation of accidents, maintenance to bid budget and assessment of overall performance every month. The company's enablers are their resources and track record. They have a significant portfolio of PFI contracts which gives them a considerable track record. They are active in almost all PFI sectors and continue to rely on the public sector to deliver more projects. The key barriers are the availability of key staff resources and the need to put equity into schemes, which stretches resources as there is a limit to the amount of money they can invest in projects.

Knowledge issues during PFI stages

Knowledge about PFI projects is sourced at the formative stage in government departments by targeting priority areas (e.g. health and education) and through discussions with consultants. There is a significant role for knowledge transfer in PFI projects, particularly for clients where there is a problem of public sector knowledge. The company regards the OBC stage as highly significant, although they are not formally involved at this stage. However, they are sometimes consulted. Table 7.4 provides a summary of the key problem areas, a description of the problem areas, staff to be targeted and scope for learning for the OBC stage.

Table 7.4   Case Study 7.2 knowledge issues at outline business case stage.

Key problem areas	Description	Staff to be targeted	Scope for learning
Unrealistic budget	Problems with matching the budget to the scope	Clients Advisers	Significant
Poor historical data	Full disclosure by client would be beneficial	Clients Advisers	Significant
Too many assumptions	Totally artificial device	Clients Advisers	Significant
Public sector comparator	Full disclosure by client beneficial	Clients Advisers	Significant

The PB stage is also considered to be highly significant as this is the stage where the two teams from the public sector and private sector are brought together into one to deliver the job. The key participants and stakeholders are clients, SPV/SPC, advisers and subcontractors who are in the front line in negotiating the details of the design. There is a need to understand sector-specific issues, PFI process, organisational delegated authority, that is knowledge of the organisation (internal Project Manager preferred). From the SPV/SPC perspective, the key activities at the PB stage are as follows:

- Detailed design to level where it is comfortable for contract signature without putting yourself at risk
- Agreement on service level (specification) and payment mechanism
- Need to finalise funding
- Due diligence – title of site, check restrictive covenants, technical survey and condition survey

Table 7.5 provides a summary of the key problem areas, a description of the problem areas, staff to be targeted for learning and scope for learning for the PB stage.

Table 7.5   Case Study 7.2 knowledge issues at PB stage.

Key problem areas	Description	Staff to be targeted	Scope for learning
Redesign scheme due to affordability	Design would have been developed over time so ought not be a problem at this stage	Clients Special purpose vehicle Advisers Subcontractors	Significant
Contractual issues	It is a challenge to get an unambiguous commitment (standard contract in education)	Clients Special purpose vehicle Advisers Subcontractors	Significant
Transfer of personnel	Unions look for certain ratios	Special purpose vehicle Subcontractors	Significant
Slow client decision-making	Clients not capable of making decisions. Need empowered people to set up a decision-making hierarchy	Clients	Significant

Case Study 7.2's main role during the FM stage is to manage the special project company. The key participants and stakeholders at this stage are FM companies, users, sector manager and bid manager. Table 7.6 provides a summary of the key problem areas, a description of the problem areas, staff to be targeted for learning and scope for learning for the FM stage.

Knowledge transfer strategy

The company has a knowledge management strategy which is related to PFI but this is not formal. There are functional heads and various knowledge sharing networks mentioned at the PB and FM stages. There are also external knowledge sharing networks as part of the supply chain which includes membership of various industry organisations active in the PFI area, meetings with particular advisers, strategic relationships with advisers, for example legal advisers and financial advisers. Reliability of knowledge from external sources and confidentiality are good as both are subject to confidential agreement.

There is a structured way of mapping knowledge but the interviewee was not aware of the details. The management board and Human Resources department constantly monitor resources to see who needs staff. Twice a year, there is a personal assessment of training needs for individuals and personal development. Learning capacity of staff is identified at the recruitment stage, and there is an appraisal process in place to monitor learning capacity.

Table 7.6   Case Study 7.2 knowledge issues at the FM stage.

Key problem areas	Description	Staff to be targeted	Scope for learning
Defects leading problems with FM	Evaluate the involvement of the FM provider earlier on	Facility Managers Users Bid Manager	Significant
Design standards and visualisation	Presentation by architects/design team will help with standards and visualisation	Clients/authority Architects Clients School governors	Significant
Managing user expectations	People tend to focus too much on design and not enough on what happens when the users start using the facility	Local authority	Significant
Change of building legislation	Results in a change in insurance requirements	Government Insurance company	Significant
Payment mechanism	FM and client requirements should be included in the contract	Client	Significant

There are mechanisms, both IT and non-IT based, to facilitate knowledge transfer, and Case Study 7.2 considers their overall readiness to transfer knowledge as high. The responsibility for knowledge transfer lies with the Financial Director and there are various resources allocated for dedicated teams, for example IT support team. The issue of organisational culture and trust is dealt with by recruiting people who are open in terms of their management style. All employees are busy, so there is no drive to hoard knowledge. However, there are no measures to monitor improvement in knowledge transfer activities.

7.4   Perspective of Consultant and Adviser

Case Study 7.3: Consultant and Adviser

Case Study 7.3 operates a global design and professional services consulting business with offices in over 100 countries. The company employs over 7000 people worldwide including professionals recognised as world authorities or innovators in their fields. Case Study 7.2 is a leading adviser, manager and designer for privately financed infrastructure projects.

Case Study 7.3's main motivation is that the organisation demonstrates strong capabilities, competencies and experience in the provision of all services required by the industry as well as the commercial rewards associated with PPP/PFI work. The key factors that determine Case Study

7.3's continued involvement in PPP/PFI are the commercial aspects of the market, the size of the market and the clients' need for consultants.

PFI/PPP strategy

The focus of Case Study 7.3's PPP/PFI strategy is on the key sectors of health, education, energy and transportation. The company has been historically strong in transportation and energy and has a dedicated management services unit, an education unit full of advisers and educations delivery specialists, and they have just acquired a health consultancy. The company intends to consolidate in the health sector, and plan to get involved in smaller schemes such as LIFT as there are few large schemes. They are also moving strongly in education and are involved in running education authorities. They are constantly looking for more work in the EU and other parts of the world such as Australia, New Zealand, South America and North America. Each sector has a think tank or team to develop a knowledge transfer strategy. In addition to dedicated PPP/PFI teams in health, education, energy and transportation, Case Study 7.3 uses virtual teams of technical specialists from various divisions across its group of companies to provide support for PPP/PFI projects. They also have a business development plan with a budget but no specific IT infrastructure solely for PFI. The company uses project extranets for handling the large amount of paperwork involved in large projects.

There are full-time staff providing PFI services operating out of nine regional offices which enable the company to provide a local presence to clients. PFI is the dominant portfolio and represents about 60–70% of their PPP activities. Knowledge about PFI projects is collected from market intelligence reports, people in public sector at the strategic planning phase, OJEU adverts, and funders and contractors from the private sector.

The company's key enabler is their knowledge of procurement and technical competencies and experience (which is their greatest asset). The main barriers are professional indemnity cover (risks) demanded by project stakeholders. They currently assess the benefits of PFI in terms of repeat work from clients, funders and request from companies to tender for work, and innovation to the market. Performance measures used to assess their performance on PFI projects are hit rates, bid costs, profitability and all the usual commercial KPI.

Knowledge issues during PFI stages

The company felt that there is a significant role for knowledge transfer. Case Study 7.3's involvement could range from the beginning to the end of the procurement process (e.g. strategic advice, business case production, OGC Gateway Reviews, advising to financial close, construction activities and/or operational monitoring).

Case Study 7.3's key activities at the OBC stage are as advisers; this consists of epidemiology planning for the health sector and project schemes for organisation and financial planning. The company has acted as Project Managers and/or advisers for NHS Trusts, local educations authorities, transportation clients and other central and local public sector authorities for preparation of OBC. Historically, the company has not been heavily involved at the OBC stage; however, this is an area where it is now expanding its business.

Table 7.7 provides a summary of the key problem areas, a description of the problem areas, staff to be targeted for learning and scope for learning for the OBC stage.

Case Study 7.3's key role at the PB stage is providing advice to clients, funders and SPV/SPC. Where Case Study 7.3's advice is sought on PFI, they normally insist on the FM provider putting a seal of approval on the design. From the perspective of a PPP/PFI consultant and adviser, the key role at this stage is providing advice to clients, funders and SPV/SPC. Their activities at the PB stage include the review of the following:

- Competencies of stakeholders
- Contractual matters
- Management systems proposed for the project

- Design (outline) used in the public sector comparator (PSC)
- Cost associated with producing design
- Town planning matters
- Construction advice (for SPV/SPC and funders)
- Operational matters for services from a consultant's point of view, output specs – key performance indicators to be achieved during the operational phase and methods statement for FM providers
- Unitary payment charge, payment mechanism (in conjunction with output specs, hard FM and soft FM)
- Financial modelling

Table 7.7 Case Study 7.3 knowledge issues at outline business case stage.

Key problem areas	Description	Staff to be targeted	Scope for learning
Unrealistic budget	The quality of the OBC is directly proportional to the budget, time available and experience of the team preparing it	Public sector organisations (e.g. Primary Care Trusts, Strategic health authorities) Advisers	Significant Some sectors have better guidance than others (e.g. health is better). Not aware of how feedback is fed into the public sector
Poor historical data	Full disclosure by client would be beneficial	Public sector organisations Primary Care Trusts, Strategic health authorities Advisers	Moderate
Too many assumptions	Totally artificial device	Public sector organisations (e.g. Primary Care Trusts, Strategic health authorities) Advisers	Moderate
Time lag between OBC and FBC (final business case)	This could lead to significant changes which negate the accuracy of the bid	Public sector organisations (e.g. Primary Care Trusts, Strategic health authorities) Advisers	Limited

How well the PB's proposals satisfy the authority's requirements and output specifications is assessed as part of the PB stage review activities. There is a need for knowledge at the PB stage about commercial awareness, contract strategy, management strategy, design, construction and cost management, and operation management.

Table 7.8 provides a summary of the key problem areas, a description of the problem areas, staff to be targeted for learning and scope for learning for the PB stage.

The FM stage is highly significant as it is the time in a project life cycle where major cost savings and efficiencies can be made. The FM and operation of a facility are the most important stage of a project; however, historically, the amount of time dedicated to designing out the issues associated with FM and operations have been minimal. PFI is addressing this issue, because in PFI projects, the FM provider now has a major input in the design stage, and as a result, buildings are now being constructed better than they use to be from an operational standpoint. Where advice is sought on

PFI, the company normally insists on the FM provider putting a seal of approval on the design. The key activities at this stage include the following:

Table 7.8   Case Study 7.3 knowledge issues at PB stage.

Key problem areas	Description	Staff to be targeted	Scope for learning
Redesign scheme due to affordability	This has a knock-on effect on the programme. If the problem cannot be solved, it is a deal breaker	Public sector organisations Advisers Special purpose vehicles Subcontractors	Significant
Contractual issues	Need to transfer risk to SPV/SPC, need to identify what risk authority could retain and understand what funders are going to accept. Go for standard/simple contracts where possible	Public sector organisation SPV/SPC Funders Advisers for: ■ Public sector ■ SPV/SPC ■ Funders	Significant
Transfer of personnel	This is down to education and public relations. For schools – no problem, but in large facilities (hospitals), there are problems with communication and trust breaks down	Public sector organisation Special purpose vehicle Facility Managers Unions	Significant
Payment mechanism	Crucial in the operational period and is a significant area to be addressed. FM involvement in design is crucial	Public sector organisation SPV/SPC Funders Advisers for: ■ Public sector ■ SPV/SPC ■ Funders	Significant
Town planning	Not advisable for any funder to invest unless planning approval granted	Public sector organisation Special purpose vehicles Local council Town planners	Limited
Construction and decant constraints	The phasing of facilities and its use is complicated because of liquidated damages issues. Issues such as dissatisfaction with finish quality and decant liquidated damages are common with refurbishment projects; therefore, it is essential to effectively manage authority expectations to be successful. Less decant the better	Design and build contractor Special purpose vehicle Public sector organisation	Significant

- Giving advice on FM on some of the potentially high-risk areas and how they should be managed.
- Independent commissioning of facilities (and checking during construction phase) and authorising payments.
- Monitoring performance through the life cycle of PFI projects, which is a new role proposed by the company which funders and clients accepted. Audit on completed PFI projects, monthly and annual inspections (in depth audit).
- Submitting reports on the performance of facilities and making recommendations to FM providers, clients and funders.

Table 7.9 provides a summary of the key problem areas, a description of the problem areas, staff to be targeted for learning and scope for learning for the FM stage.

Table 7.9   Case Study 7.3 knowledge issues at FM stage.

Key problem areas	Description	Staff to be targeted	Scope for learning
Managing user expectations	Need for a change in client's attitude. Need for a cultural change in staff transferred from the public sector. Need to re-educate commercial organisations	Facility Managers Clients	Significant
Defects leading to problems with FM (interface agreement)	Not had any problem but that is not to say it would not happen in future. If clerk of works is appointed by FM team, then monitoring will be in the interest of FM provider	Facility Managers Design and build contractor Funders Technical advisers Clients	Significant
Design standards and visualisation	Not yet come across any facility that has failed in design terms. FM provider will pick it up where there is a requirement for audit	Facility Managers Design and build contractor Funders Technical advisers Clients	Significant
Change of building legislation	Need for agreement on how change could be absorbed	Clients Government Professional bodies EU	Significant
Staff transfer	Issue with staff to be transferred to private sector	Public sector organisations Facility Managers Unions	Significant
Payment mechanism	After years of operation, many were proved to be incorrect	Facility Managers Design and Build Contractor Funders Technical advisers Clients	Significant

Knowledge transfer strategy

Case Study 7.3's knowledge transfer activities are group related and not specific to PFI. The company has a 10-year strategic plan to embed a KM culture across the group, professionally – 'a learning culture, continuously learning, listen and then able to lead'. Professional excellence and KM were brought forward. The company believes that their readiness for KM to be good as they are already in the implementation stage led by a KM champion in the main board appointed part-time since 2003. Resources include a budget and KM steering group (as it is not the company's culture to separate people from projects), and people are tasked to allocate a proportion of their time.

A number of transfer mechanisms are adopted. Some of these are as follows:

- Discussion groups initiated by staff members where issues resolution and innovative solutions are driven from the bottom-up
- Knowledge sharing networks within the organisation
- Project extranets to link external customers
- Knowledge mapping
- Technical forums
- Communities of practice
- Improved communication facilities (e.g. conferencing facilities, white boards, yellow pages, shared calendars for setting up meetings, etc.)

The company is keen to develop a knowledge transfer strategy particularly for the OBC stage to strengthen their role as advisers and consultants as this stage is most critical for clients. However, existing internal culture have created some resistance to this. There is also a need to set up systems to formulate asset registers and computerise everything with respect to FM operations. Workshops are also needed to educate people on how to use asset registers, maintenance systems and performance audits to identify things that are wrong and feedback to computerised maintenance systems to ensure that appropriate planned preventive maintenance is performed. When measuring performance for FM providers, it is best to measure the calls that the 'help desk' receives where each fault is recorded. Improved performance is evidenced by a reduction in the number and severity of faults reported. There is considerable scope for an FM knowledge transfer strategy as it will benefit everyone (including clients, current and future projects). The training of younger consultants to rapidly build core competencies in FM and operations is a priority for the company. The company argued that the best way to learn is 'hands-on experience' and knowledge transfer workshops that address real-life issues. A formal process for measuring competencies exists using an in-house PFI training tool which asks a series of questions on PFI processes to identify training needs. However, due to resource constraints, it is not applied within the organisation as it should be. The company agrees that there is a need for a systematic knowledge transfer strategy; however, resource constraints are an issue.

Following the knowledge mapping initiative, knowledge and expertise are now more widely dispersed to include centres of excellence in the United States and a part of Asia. There are plenty of measures in terms of the intranet including discussion groups.

7.5   Perspective of Design and Build Contractor and FM Provider

Case Study 7.4: Design and Build Contractor and FM Provider

Case Study 7.4 is one of the world's leading companies in the project management and construction services industry. It operates in about 40 countries spanning six continents and employs over 10 000 people. About a third of the company's current and past clients are in the Fortune 500 list of the world's leading businesses. The company is involved in high-profile PFI projects.

Case Study 7.4's motivation for participating in PPP/PFI is driven by the long-term nature of the projects, profitability, limited competition, huge projects – the type of projects they are able to carry out. It brings together different parts of the group's activities from development, construction, FM and investment. There is a very large market with a deal flow that allows them to specialise in the health care sector. Its balance sheet means that they can undertake the largest more complex health PFI. High risk means higher returns, attractive for creating opportunities for profit over a long time.

PFI/PPP strategy

There is an overall PPP strategy for Case Study 7.4's owners. In the United States, they are involved in military towns only, community infrastructure in Australia, in Europe (Italy and Spain) and in the United Kingdom (PFI). They focus on specific sectors, differentiating between sectors, and are very strategic in 'where we go' to use the management company as the integrator. The company's continuous involvement in PFI/PPP projects depends on making a good return on the risks involved, the market remaining sustainable, maintaining a good win rate because of the high cost of bidding. 'It is more economics rather than political' as both political parties have embraced it. They are consolidating in health sector as they recognised that they must be amongst the top. Education was not showing enough returns but BSF schemes could be attractive. They are involved in Ministry of Defence projects and might consider going back into prison PFI projects. Every sector competes for capital.

Different parts of Case Study 7.4 are responsible for different areas. There is a PFI strategy which relies on core bid teams, significant resources (budget), FM teams, general management, health and safety and finance teams. For each new project, a consortium is formed and investors are invited. The company is responsible for managing each of the SPC (special purpose companies). They have FM partners in Europe such as France and Denmark. There is no specific infrastructure but they rely on support provided by the group's fully integrated system. For example, there is a 3D system in development to facilitate asset management.

PFI is the most significant of Case Study 7.4's PPP portfolio. They adopted a sector approach; they keep people in the health sector once they are there. The company moves people to different functions so that employees have a good knowledge of a number of areas to feed this back to front-end design. Intelligence about PFI is developed through talking directly to National Health Trust, financial advisers, their own network of information and constant liaison with the Department of Health's PFI unit.

They currently assess the benefits of participating in PFI/PPP in terms of superior returns (profitability), changing the culture of construction business (skills set change) as PFI is beneficial to the skill culture of mainstream construction. Each element of PFI activity is measured for investment returns, each project is assessed for risks and margins, delivery to time and budget, and safety which is crucial measure for performance and reputation.

The key enablers (internally) are having the right skills set, right place and time as PFI is very resource intensive requiring about 150 design staff at peak design workload. Externally, the skills we do not have: for example health care planners and health care architects. The key barriers are bid capital – losing means writing off millions in bidding for no result, risk profile and the context of the project (planning, logistics, politics, etc.).

Knowledge issues during PFI stages

Case Study 7.4 considers the OBC stage as important; they tend to follow the brief and will research the OBC to make sure it is sound but their role is limited or non-existent at this stage, the OBC tends to lack the correct solution/model for care. What is required at this stage is enough details to get outline planning permission. The company's role is limited or non-existent at this stage but they have health planning specialists and architects heavily involved in the OBC. The key participants are the client project team, NHS advisers who approve the OBC and who gave the guidance, town planners, legal, technical advisers (architects, quantity surveyors, engineering specialists) and financial advisers. It is important to draw on the experience of clinicians and administrators including health care planners and estate advisers. There are particular types of knowledge required such as good project management ability and technical skills to set the standards to be met by the OBC (not cut and paste from the previous projects), good surveys and good technical advice, enough access to technical advice as some NHS Trusts do not want to spend money on specialist advice such as acoustics and traffic, and so on. Table 7.10 provides a summary of the key problem areas, a description of the problem areas, staff to be targeted for learning and scope for learning for the OBC stage.

Table 7.10   Case Study 7.4 knowledge issues at outline business case stage.

Key problem areas	Description	Staff to be targeted	Scope for learning
Unrealistic budget	Identify early and insist clients have early affordability meeting. If clients are responsive, then negotiations can proceed	Client project team NHS advisers Town planners Legal advisers Technical advisers Financial advisers	Significant
Poor historical data	Get earlier surveys. Work with other competitors	Client project team NHS advisers Town planners Legal advisers Technical advisers Financial advisers	Moderate
Too many assumptions	Allow contractors to amend assumptions	Client project team NHS advisers Town planners Legal advisers Technical advisers Financial advisers	Significant
Too much detail in OBC	Time is wasted in this	Client project team NHS advisers Town planners Legal advisers Technical advisers Financial advisers	Significant

The PB stage is highly significant for Case Study 7.4. By winning the bid, the company secures the work but they do not recover the bid cost until they have reached financial close. Key activities involves at this stage are as follows:

On the technical side:

  ▪ Signing off of 1/200, 1/50 drawings with client directors (develop 1/200 and 1/50 drawings to get them signed off). Check that every room has everything that is needed, for example fixtures.

- Prepare schedule 8 (largest schedule in the contract) to put the full business case (FBC) to the NHS.
- Trust needs schedule 8 to the FBC.
- Set of drawings (1/200) to get full planning permission.
- Work out other technical details, performance standards, energy levels and equipment.
- Design build FM issues (about how building will operate).
- Schedule 8 will be encapsulated in the contract which will be signed at financial close (FC). Other schedules are 14 and 18, service level specification and payment mechanism.

On the commercial and legal side:

- Negotiate contract and price (standard contract SFPA3). Benchmark for negotiation is the bid price with underlying assumptions (about risks, etc.).

Table 7.11 provides a summary of the key problem areas, a description of the problem areas, staff to be targeted for learning and scope for learning for the PB stage.

Table 7.11   Case Study 7.4 knowledge issues at PB stage.

Key problem areas	Description	Staff to be targeted	Scope for learning
Redesign scheme due to affordability	Prefer no redesign, look at options and present to client	Design management team Design and build contractors Client Both financial advisers	Significant
Contractual issues	Standard forms	Legal advisers on both sides Client Special purpose vehicle Facilities Manager	Limited
Transfer of personnel	FM providers experience in TUPE[a] transfer	Facilities Manager Unions Clients Special purpose vehicle	Significant
Clients not geared up for FC on time	Either do not have the resource or experience to close deals or their inability to manage stakeholders	Special purpose vehicle Client Legal advisers Financial advisers Subcontractors	Significant

[a]Transfer of Undertakings (Protection of Employment) Regulations.

The FM stage is also highly significant. That is where the majority of the risks are, unlike traditional contracting. Problems tend to manifest itself in the operational stage. Knowledge comes from three main sources: lessons learnt, estates manager (traditionally not part of the construction team) dealing with the day-to-day running or operation of the facilities and construction teams who have built hospitals before. FM is split into soft services and hard services. Case Study 7.4 generally partners with a specialist FM company on the soft side (responsible for e.g. porter and security, linen and catering (areas for retention of employment)). The company handles the hard services which include estate maintenance, grounds and gardens, pest control, and so on. The FM model is to do with the history and core business of the company which is construction, hence hard services. Both soft and hard services are important as they operate under financial pressures and penalties. The risks are more aligned to the hard estate services (e.g. power outage, water supply problems, infection control). The level of risk is different in the soft side. However, infection control sits at the interface between hard and soft FM. The NHS has an infection control officer now a director who sits on the board in recognition of its importance. Estate services depend on how the design is put together, keeping the building performing to environmental standards. Soft side deals with people support – it is people intensive (e.g. 1000 compared to 100 for hard side). Earlier PFI models anticipated that government will divest immediately of the people and everything will be transferred to the private sector. However, items and conditions are protected under retention.

Table 7.12 provides a summary of the key problem areas, a description of the problem areas, staff to be targeted for learning and scope for learning for the FM stage.

Knowledge transfer strategy

There is a role for knowledge transfer. For Case Study 7.4, it is the most profound requirement after working capital. The company has a knowledge management strategy but not specific to PFI. They consider themselves very poor at sharing knowledge. There are informal knowledge sharing networks within the organisation but they are trying to develop formal ones. There are knowledge sharing networks for specific contractual issues, for example JCT or PFI. There are also informal knowledge sharing networks with the supply. There is no method for mapping knowledge. Training needs are identified through appraisal and project plans, and in the process, learning capacity is assessed.

There are mechanisms, both IT and non-IT, to facilitate knowledge transfer. The company has a knowledge bank software, and its meetings and informal networks are used to transfer knowledge. There is an informal role for knowledge sharing which is the responsibility of the Technical Director and one of the business improvement managers. There is no specific budget or staff allocated for knowledge management activities as this is expected to be part of employees' normal work. The issues of organisational culture and trust are being addressed top–down. The company considers themselves as a team organisation and team organisation 'should be sharing knowledge'. There are no performance measures to monitor improvement in knowledge transfer activities.

The mechanisms at the PB stage are Communities of Practice (CoP) in health care, feedback from IT knowledge sharing system, lessons learnt on previous projects and feedback from Trust to give the project team impartial advice. They do not have measures to assess the mechanisms for improving learning. Learning mechanisms are the same with the PB stage but there is also a design standard for FM guide – best practice document. Usually, they can tell when somebody has read the design standard for FM guide but do not have measures for learning. From an FM point of view, there are a lot of measures – humidity, lifts and operating theatre from the monthly report. Lifts are monitored continuously and a graph is plotted. Lift manufacturers also monitor remotely as somebody might require help if stuck in the lift (requirements of legislation). There is scope for a knowledge transfer strategy to help deal with the issues.

7.6   Key Problem Areas and Scope for Learning

All four case study companies expressed concerns about knowledge-related issues across the key stages of OBC, PB and FM. The key problem areas were common to the stakeholders. The following section highlights the key points arising from the interviews in terms of problem areas and scope for learning.

Table 7.12   Case Study 7.4 knowledge issues at the FM stage.

Key problem areas	Description	Staff to be targeted	Scope for learning
Defects leading to problems with FM	Not spending enough money upfront to avoid problems in the future. Interface agreement exists but cannot cover everything, it is an intent	Facility Managers Design and build contractors	Significant
Design standards and visualisation	Design standards do not reflect what an FM operator wants	Facility Managers Design and build contractors	Significant
Managing user expectations	The NHS Trust in-house expectations are different from reality of what the funding can provide	Client	Limited
Change of building legislation	There are mechanisms to address the problems in FM contract	Previously Client and Facility Managers (old contract) Facility Managers (new contract)	Limited
Infection control	Biggest problem is infection control – water supply, pigeons affecting air intakes, water storage capacity. Most important is the Trust. Problems often manifest itself in the hard services (water supply – adverse publicity in estate services)	Client Facility Managers (soft and hard FM)	Substantial
Penalties	Second biggest problem are the penalties. In the new model, penalties are stiff and margins are slim. More scope to make profit on the soft FM side. Risks are on all the FM side. Trusts set the charges unilaterally, but SPV/SPC tries to influence the Trust in this area. Two types: (a) non-availability of services, and (b) failure events. Payment is dealt with using the upward chain – soft and hard FM provider generate a valuation of their performance which is submitted to SPV/SPC (who would then verify) and forward to the Trust to justify payment. A downward chain is followed if there is a problem noticed by the Trust	Client SPV/SPC Facilities Manager Funders	Limited

7.6.1   Outline business case

Seven main areas were identified as requiring better transfer of knowledge. These were as follows:

- Unrealistic budgets
- Poor historical data
- Too many assumptions
- Confidence in the financial data
- Public Sector Comparator
- Time lag between OBC and PB stages
- Too much detail in the OBC

All stakeholders identified the first three points highlighting the need to provide knowledge for the preparation of the financial aspects of the OBC stage. One of the problems lies in trying to make the project scope fit the budgetary requirements. It is also interesting to note that the client is the one stakeholder least concerned with learning about the financial implications of decisions made at the OBC stage, preferring to pass this responsibility onto external financial advisers. The other stakeholders are very keen to transfer learning from this stage to produce more realistic budgets because they are the parties that will have to live with the decisions made at this stage. A number of learning mechanisms are exploited to improve knowledge transfer at this stage. These include the following:

- Using internal networks, for example CoP and intranets
- Using external networks and advisers to help share knowledge, for example 4Ps
- Attending seminars, workshops and conferences
- Speaking to the client and reading technical papers issued by the client
- Gaining access to previous bids and examples of work

Thus, the main tools required for this consist of creating knowledge of financial data and having access to relevant data and making it accessible to the stakeholders.

7.6.2   Preferred Bidder

Nine key problems areas were identified at the PB stage. These were as follows:

- Redesign scheme due or affordability
- Contractual issues
- Transfer of personnel
- Land assembly
- Slow client decision-making
- Payment mechanism
- Town planning
- Construction and decant mechanism
- Clients not geared up for FC

Like the OBC stage, the first three points were highlighted by all case study companies. The impact of redesigning to match the budget and the knock-on effect on the construction programme is a big issue. The problem with lack of contractual knowledge amongst all stakeholders is identified. This is a particularly sensitive area with a 25-year contractual obligation presenting additional challenges and an unwillingness to accept risk. PFI also puts pressure on staff. With so few people having expertise in PFI, employees are continuously being transferred to bids where their expertise is at a premium.

Again, the client appears to be the one stakeholder out of step with the other stakeholders in terms of the scope for learning. The client has rated the scope for learning as moderate whereas the other stakeholders rate this as significant. To address the above, a variety of learning mechanisms can be deployed to improve knowledge transfer. These include the following:

- Using meetings (e.g. with bid team managers) and presentations (both from internal staff and from external specialist advisers)
- Using internal networks (e.g. CoP) and external networks (e.g. 4Ps)
- Using IT services such as EDMS and the company intranet
- Adopting lessons learned from previous projects
- Obtaining feedback from clients

7.6.3   Facilities Management and Operational Phase

The stage resulted in seven areas identified as benefiting from knowledge transfer. These include the following areas:

- Defects leading to problems with FM
- Design standards and visualisation
- Managing user expectations
- Change of building legislation
- Infection control
- Payment mechanism
- Staff transfers

The FM category obtained more consensus amongst all stakeholders agreeing with the first four points. The problems mainly concern lack of communication between various stakeholders such as not exploiting the expertise of FM during the design or insufficient communication between the technical experts and the building's users. In this case, knowledge transfer between the groups would alleviate some of the problems. Other problems such as changing building legislation and payment mechanism provide opportunities for knowledge transfer.

Again, the attitude of the client organisation is that their scope for learning during this phase is minimal whereas other parties/stakeholders consider this a critical area for learning, mainly because decisions made in earlier PFI stages have such a profound impact on the FM stage.

Specific learning mechanisms identified to improve learning during the FM stage include the following:

- Using internal networks (e.g. CoP) and informal meetings
- Using external networks and advisers to help share knowledge (e.g. 4Ps)
- Exploiting IT to gain access to key knowledge from EDMS and Skills Yellow Pages and corporate intranets
- Developing an asset register and develop performance audits to identify maintenance issues
- Training of younger consultants to build core competencies

7.7   The Need for Knowledge Transfer

The discussion in the previous section highlights the key problem areas identified during the OBC, PB and FM stages, which can be divided into two categories. First, those problems caused by a lack of knowledge to inform decision-making, specifically financially based decisions required during the OBC and PB stages. Second, those decisions that can be better informed through improved knowledge transfer amongst the project stakeholders.

The first category, the problems that arise from a lack of knowledge, can be addressed by 'creating' knowledge to solve those problem areas. PFI is a relatively new form of procurement without a long track record of project data. What these case studies show is that there is an urgent need for project benchmarking data that can be used to better inform decisions made at the early PFI stages. This calls for knowledge elicitation from the few PFI experts and a codification of this knowledge. This knowledge can be elicited from a wide range of completed and current PFI projects, covering all the key PFI sectors and stages. It is strongly recommended that a central body such as 4Ps creates and maintains this benchmarking data so that new PFI project stakeholders can exploit the knowledge and expertise derived from past projects. This is also an area where IT can play an important role in indexing, storing and disseminating the knowledge created.

The second category, that of the transfer of knowledge between various stakeholders, is required for both intra-project (between the various PFI project stages) and inter-project (between PFI projects) knowledge transfer (Kamara *et al.*, 2002). One of the key issues here is fostering better communication and making communication easier between the various stakeholders. The project participants need to have access to the data required whether from internal sources (e.g. other departments and business units) or externally (e.g. from advisers). Unlike the first category, knowledge to solve the key problem areas does exist but this knowledge has not been transferred to those that need it, at the time they need it. The project stakeholders therefore need to identify what knowledge is required, who holds that knowledge and develop knowledge transfer mechanisms to acquire that knowledge. This area will be elaborated in Chapter 9. Communication is at the heart of the problem and therefore project stakeholders need to identify both IT and non-IT tools that will enhance communication (Al-Ghassani *et al.*, 2002). IT tools such as

Skills Yellow Pages and project extranets can be adopted to identify specific types of PFI experts. Non-IT tools such as seminars, workshops, technical meetings, professional networks, best practice guides, project reviews, and so on, can be exploited to transfer the knowledge required.

It is envisioned that with the increasing number of PFI/PPP projects, benchmarking data will become more readily available and greater communication between project stakeholders will alleviate some of the problems highlighted. However, in key PFI stages highlighted in the case studies, the client was self-assessed as requiring less scope for learning than the other stakeholders. This implies a greater reliance on other team members either because of a lack of in-house resources or a failure to recognise that the client organisation needs to improve their learning as much as any other stakeholder to make PFI projects a success. It is crucial that client organisations recognise that they need to be as skilled in PFI/PPP procurement as the private sector if they are to take a leading role. Knowledge transfer should therefore form a key component of client organisations' strategy towards PFI.

PFI projects are relatively new to the industry and their unique procurement method means that there is a lot of scope for transferring knowledge both between projects and within individual project stages. The case studies clearly indicate that there are issues regarding knowledge at various stages of the PFI process that are common to different stakeholders. Some of the issues are concerned with creating a knowledge base (e.g. lack of historical data), and others clearly have a knowledge transfer element (e.g. issues concerning contractual issues).

Some of the issues highlighted such as the creation of unrealistic budgets identify a need for knowledge creation and other issues highlight the need for knowledge transfer both within the case study companies and between case study companies. In order to improve knowledge transfer, a systematic approach should be adopted. Each case study company needs to identify precisely what knowledge needs to be transferred, the characteristics of the knowledge that need to be transferred and then determine, based on a wide range of available tools, the best mechanism for transferring that knowledge.

7.8   Improvement Capability and Organisational Readiness

The case studies show that public sector client organisations clearly need to improve or show a willingness to improve. Private sector organisations appear much more willing to improve to win new work. Although there are opportunities to share or transfer knowledge within and between organisations on long-term projects such as PFI/PPP projects with multiple stakeholder, there are a number of factors that can affect a knowledge transfer strategy such as improvement capability or an organisation's ability to learn and absorb new knowledge (Gann, 2001). Organisational readiness or willingness to share particular types of knowledge or change its culture to facilitate knowledge transfer activities is also crucial (Crawley and Karim, 1995; Bresnen and Marshall, 2000).

There are measures from the private sector perspective to assess the benefits of PFI/PPP projects such as profitability/returns, repeat work from clients/funders and bid success rates. From a public sector client's perspective, measures include improvement in pass rates, pupil's attendance, staff retention, morale and absenteeism for schools, safer roads and better street lighting for transport PFI/PPP projects. However, most of the construction and client organisations noted that there is a scope for learning and improvement in key issues through learning and knowledge transfer (e.g. budgeting, historical data and assumptions) associated with the OBC stage and other key issues at the PB and the FM stages to improve practices and expand business opportunities. Whilst it is recognised that different types of knowledge are required and a variety of tools are used for learning and knowledge transfer activities, measuring learning and the extent of knowledge transfer to assess effectiveness remains a major problem which has to be addressed to improve an organisation's ability to learn or its capability to absorb new knowledge.

There are a number of key enablers for participating in PFI/PPP projects such as having adequate resources and staffing, track record and, from a client perspective, getting elected members/councillors, schools and governors support in a case of local authorities. However, there are number of barriers associated with an organisation's ability or readiness to effectively implement knowledge transfer activities such as organisational culture creating a resistance to knowledge transfer, various resources required such as personnel and IT infrastructure to support knowledge transfer activities and the need to monitor knowledge transfer activities.

7.9   Concluding Remarks

This chapter investigated the experiences of four case study companies on PFI projects. The case study companies provide an insight into the knowledge issues experienced by a range of stakeholders. The implementation of PFI/PPP projects is often regarded as a way to achieve a step change in improving services and assets, but there are problems in key phases and stages in the delivery process. What is interesting is that several companies identified the same sort of problems. Many of the issues identified concern the creation and transfer of knowledge. However, before knowledge can be transferred, firstly, there must be a willingness to acknowledge that there is scope to learn on specific issues; secondly, there must be an acknowledgement that knowledge transfer is required. The knowledge transfer strategy and mapping of most organisations are at best sketchy, inadequate and informal. There is therefore a need for a knowledge strategy to improve business opportunities underpinned by appropriate knowledge mapping techniques, an action plan for learning and capacity building activities with measures to assess the effectiveness of learning, knowledge transfer tools and organisational readiness to facilitate continuous improvement in the delivery of PFI/PPP projects. The next chapter examines a framework that will help organisations to identify how best to transfer knowledge in order to learn and thereby address some of the key knowledge issues.

References

Al-Ghassani, A.M., Robinson, H.S., Carrillo, P.M., and Anumba, C.J. (2002) *A Framework for Selecting Knowledge Management Tools*. Proceedings of the 3rd European Conference on Knowledge Management (ECKM). Dublin, Republic of Ireland, pp. 37–48.

Bresnen, M., and Marshall, N. (2000) Partnering in construction: a critical review of issues, problems and dilemmas. *Construction Management and Economics* 18(2), 227–237.

Crawley, L.G., and Karim, A. (1995) Conceptual model of partnering. *ASCE Journal of Management in Engineering* 11(5), 33–39.

Gann, D. (2001) Putting academic ideas into practice: technological progress and the absorptive capacity of construction organisations. *Construction Management and Economics* 19(3), 321–330.

Kamara, J.M., Anumba, C.J., and Carrillo, P.M. (2002) A CLEVER approach to selecting a knowledge management strategy. *International Journal of Project Management* 20(3), 205–211.

Knowledge and Capacity Building Challenges

8.1 Introduction

The case studies in the previous chapter highlighted some of the knowledge transfer and capacity building issues faced by different stakeholders such as clients, special purpose vehicle (SPV/SPC), design and build contractors, facilities management providers, designers and advisers in PFI/PPP projects. Firstly, it is important to understand the nature of these challenges, and secondly, determine how knowledge management can address specific problems by identifying the mechanisms to stimulate the development of tacit and explicit knowledge for organisations, institutional and national capacity building.

This chapter focuses on the role of transfer capacity building in improving performance of PFI/PPP projects. It starts with a discussion of some of the key issues affecting PFI/PPP projects based on the findings from a survey carried out to (1) identify the level of participation and different roles, motivation and perception, (2) review current practices in procurement/planning, construction and the operation and service delivery phases and (3) investigate barriers and enablers in the delivery of PFI/PPP projects. The scope for learning, lessons learned and the implications for capacity building in PFI/PPP projects are also examined. The chapter discusses various options in capacity building to develop tacit and explicit knowledge through the creation of knowledge networks and communities of practice, strengthening of knowledge centres, the development of dedicated PPP units and support agencies. The role of guidance documents and dissemination of best practices to share lessons learnt and to improve PFI/PPP processes are also examined. The creation of knowledge and transfer of tacit knowledge through traditional training and learning approaches, research and development, external advisers and the role of technical assistance in capacity building are also discussed.

8.2 Roles and Participation

One of the largest and earliest questionnaire survey carried out on PFI/PPP issues was conducted by Robinson *et al.* (2004). The survey received responses from 100 public sector client and private sector organisations in the United Kingdom out of 173 questionnaires giving an overall response rate of 58%. The public clients include NHS Trusts involved in health care projects and local authorities involved in education, transport and other types of PFI/PPP projects. The private sector includes consulting firms and contractors with some organisations having multiple roles in PFI/PPP projects such as SPV/SPC, design and build subcontractors, designers, advisers and facilities management subcontractors.

About a quarter (25.5%) of construction organisations have their main role as SPV/SPC, 23.4% as design and build contractors, 21% as designers, 17% as advisers and 6.4% as service providers/operators. However, there are significant differences between consulting firms and contractors not only in terms of their primary role but also in the level of participation. About a third (32%) of consulting firms act as advisers to client organisations and funders/lenders and another 40% operate as designers who are part of the design and build team or SPV/SPC. In contrast, many of the contractors operate as SPV/SPC (38.1%) and under half (47.6%) as design and build contractors as their primary role. About a tenth of contractors (9.5%) are involved as facilities management service providers/operators. Out of the organisations contacted initially, 49 out of 57 consulting firms were actively involved in PFI compared to 37 out of 64 contractors, giving a participation rate of 86% and 59% for consultants and contractors, respectively.

8.3   Motivation and Perception

PFI/PPP projects offer tremendous opportunities for a wide range of public sector and private sector organisations in the construction industry. The motivation for participating and the views of these organisations on PFI projects were investigated relating to costs of PFI projects, innovation, risk and value for money. The survey identified that public sector clients and private sector construction organisations have different reasons for participating in PFI projects.

Client or public sector organisations identified government policy, and the requirement not to provide capital funding as key drivers. They are involved in PFI/PPP projects as it is seen as a way to access significant funding for large public infrastructure projects that would otherwise not be funded from public budget using traditional procurement due to financial constraints. Public sector client organisations will continue to be involved in PFI/PPP projects as long as they can engage the private sector and demonstrate best value for money. However, private sector construction organisations are motivated mainly by the steady, long-term income stream and diversified workload as traditional construction activities are cyclical with major troughs and peaks. Higher returns and profitability are also the key drivers and their continuous involvement depends on maintaining good commercial returns, effective bid management to reduce bid and transaction costs and high level of political commitment to ensure the continuous flow of PFI/PPP deals. Figure 8.1 shows the most important motivating factors selected by the 48 public sector clients and 52 private sector construction organisations.

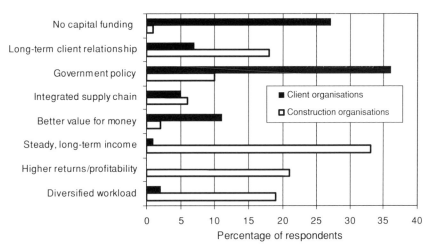

Figure 8.1 Motivation for involvement in PFI/PPP projects.

8.4   Value for Money, Costs, Innovation and Risks

Value for money and risks transfer are key issues in PFI/PPP projects. However, the value for money obtained on PFI/PPP projects has often been questioned in relation to the cost of traditional forms of procurement. Table 8.1 shows that the majority of public sector clients and private sector construction organisations believe that the bidding costs of PFI are higher, particularly for smaller and medium-sized projects with responses of 88% and 71%, respectively. Only about a third believes that the costs are higher for the design and construction phases. The high bidding costs are due to the complexity of the process, associated costs for specialist legal, technical and financial expertise required and the transaction costs during the lengthy negotiation periods for PFI/PPP projects.

Table 8.1 Proportions that consider PFI costs to be higher than traditional procurement.

	Number of client organisations (out of 48)	Number of construction organisations (out of 52)	% Total respondents
Bidding costs			
Small projects (<£30 million)	42	46	88
Medium projects (£30–£70 million)	33	40	71
Large projects (>£70 million)	26	37	63
Design and construction costs			
Small projects (<£30 million)	16	23	39
Medium projects (£30–£70 million)	14	20	34
Large projects (>£70 million)	11	19	30

£30 million = US$ 57 million; £70 million = US$ 133 million at the time.

PFI projects were expected to bring about increased innovation (due to the consortia created), transfer risk to the private sector and provide value for money for the government. Davies (2006) argued that by internalising 'project maintenance costs post-construction, PFI contractors may have an incentive to install more efficient types of technology and deliver the project at a lower cost'. The lower cost from the PFI consortium can be achieved due to the strong incentives to 'reduce costs but not to jeopardise quality', innovation and better risk management practices from the private sector. A key benefit is the opportunity for innovation in terms of whole life approach and the integration or synergy between design, construction and asset management functions to enhance the delivery of services. Innovation can result in significant operational cost savings (Ball *et al.*, 2000). However, innovation in design and service delivery and its impact on whole life costs and value for money is often the subject of intense debate. de Lemos *et al.* (2003) found that whilst designers are free to innovate in their designs, contractors were conservative in the materials they used. Also, certain sectors, such as health, left little room for innovation because of the standards required. The results of the survey showed the following:

- Fifty-four per cent of respondents considered PFI to produce improved innovation in design.
- Fifty-two per cent considered risks and rewards were appropriately managed.
- Fifty-two per cent considered PFI provided value for money for the whole life performance.

These figures show that neither public sector client nor private sector construction organisations are completely convinced that PFI/PPP projects deliver the anticipated benefits.

8.5   Enablers and Barriers

It is important to understand the key enablers and barriers affecting the delivery of PFI projects to facilitate continuous improvement. The Treasury acknowledges that there is room for improvement, particularly in the areas of reducing the procurement timescales and reducing procurement costs (HM Treasury, 2003). In this regard, a number of enablers and barriers are considered important such as PFI expertise, procurement periods, complexity of bidding, transaction costs and other unique PFI issues.

8.5.1   Expertise and knowledge

A key enabler is the level of PFI knowledge in the public and private sector organisations. PFI is a relatively new form of procurement with many organisations vying for staff with PFI expertise. This is much more prevalent in public sector client organisations because few have undertaken more than one PFI project and therefore have to rely heavily on technical, financial and legal advisers. A National Audit Office report (NAO, 2001) identified the need for local authorities to have the right skills to manage PFI projects and the problems with staff continuity on PFI projects. The survey results showed an average experience of construction organisations in PFI projects is 7.3 years compared to 5.4 years for client organisations. *Private sector* construction organisations have more experienced staff compared to public sector client organisations. Overall, about 73% of construction organisations rated their company's expertise in PFI to be 'good' or 'very good', compared with 64% of client organisations. Just under a quarter of organisations rated their expertise as 'satisfactory' whilst no organisation rated their expertise as 'very poor' or 'poor'.

Lack of PPP/PFI knowledge and expertise can undermine the implementation of PPP/PFI programmes. There is some evidence to suggest that the limited PFI/PPP contractors in the United Kingdom specialising in complex hospital projects threatened to undermine the level of competition required to achieve value for money in PPP/PFI projects. Economists argue that even if a country has succeeded in raising investments, it takes decades to transform master plans, programmes and projects into infrastructure capital (i.e. roads and highways, power plants, hospitals, rail networks, factories, water supply systems) that underpins a productive economic structure. This is often because of skills and capacity constraints. There is a range of expertise required and lack of skilled personnel and other resources identified in the policy framework can affect the implementation of PPP/PFI projects.

8.5.2   Procurement periods

The time taken for PFI projects is a growing concern as it has a significant influence on bidding and transaction costs. Little seems to have changed since Ezulike *et al.* (1999) highlighted the problem of the extensive time required for bidding. Table 8.2 shows the average timescale in months from preliminary invitation to negotiate (PITN) to preferred bidder (PB) and financial close (FC) under the negotiated procedure. Defence PFI projects were a

Table 8.2 Procurement periods by sectors (in months).

	Health	Education	Transport	Custodial	Defence
Client organisations					
PITN to PB (a)	13.0	13.2	*	*	*
PB to Stage FC (b)	13.7	6.3	*	*	*
PITN to FC (a + b)	26.7	19.5	*	*	*
Construction organisations					
PITN to PB (a)	12.0	10.2	12.7	16.7	16.5
PB to FC (b)	11.2	9.3	8.2	11.2	18.0
PITN to FC (a + b)	23.2	19.5	20.9	27.9	34.5

* means insufficient data/data not available.

major problem requiring an average time of 34.5 months, well above the other sectors. Both the public sector client and private sector construction organisations agree that the procurement time is too long. Sixty-three per cent of client organisations and 80% of construction organisations believe that the time lapse between PITN and PB is too long. Similarly, 65% of public sector client organisations and 75% of private sector construction organisations believe that the period from PB to FC is too long. The implications are that both private sector construction and public sector client organisations have staff involved/committed in PFI projects for lengthy periods, thus tying up vital or scarce resources without a known outcome.

8.5.3   Other barriers

A number of other barriers also affect the participation of organisations in PFI/PPP projects. Figure 8.2 shows the most important barriers. High transaction and bidding costs associated with the procurement periods are a major issue with some failed/aborted bids incurring more than £1 million which then have to be recovered from other projects. Ball *et al.* (2000) acknowledged that the high process costs associated with bidding could make PFI projects more expensive than traditional projects which can undermine the value-for-money argument.

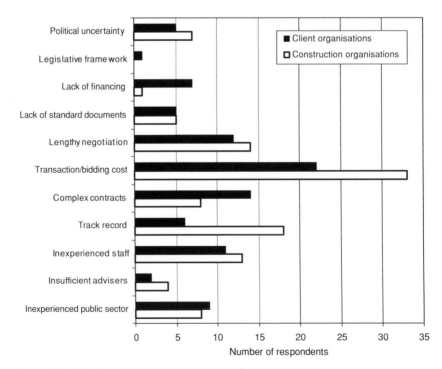

Figure 8.2   Barriers to participation in private finance initiative.

The most significant barriers identified by public sector client organisations, in order of importance, are:

- high transaction and bidding cost,
- complex contracts,
- lengthy negotiation periods.

For private sector construction organisations, the most significant barriers identified are:

- the high transaction and bidding costs associated with PFI,
- lack of experience with PFI,
- lengthy negotiation periods,
- inexperienced staff.

8.5.4   Unique PFI issues

There are a number of unique issues associated with PFI at the planning and design development phase, sometimes called the 'procurement' phase, construction and operational and service delivery phases that are of growing concern to public sector client and private sector construction organisations. Table 8.3 shows a summary of respondents' perception or average ratings of key issues relating to the three phases of PFI projects. The issues of main concern are those with higher average ratings based on a scale from not significant (1) to highly significant (5).

In general, public sector client organisations tend to experience far more problems with PFI/PPP projects than private sector construction

Table 8.3 Unique private finance initiative issues at key phases.

	Average score – public client organisations	Average score – private construction organisations
Planning and design development phase		
Inadequate client brief/requirements	2.7	3.6
Lack of time given to bidders	2.0	2.9
Lack of staff resources	3.5	2.7
Difficulties establishing life cycle costs	3.1	2.3
Affordability/funding gap	4.5	3.8
Poor project management	2.4	3.2
Construction phase		
Design change orders/variations	3.5	3.3
Defects and rework	3.1	2.4
Cost overrun	2.3	2.6
Construction delay	2.8	2.6
Service delivery and operation phase		
Difficulties in sustaining service level	3.1	2.4
Difficulties in maintaining facilities	2.5	2.3
Remuneration and payment dispute	2.5	2.3

organisations. In the planning and design development phase, affordability and funding gap is very highly rated by client organisations with almost all of the respondents (92%) agreeing that these are very significant and structural issues at the heart of PFI projects. Also, lack of staff resources is more of a problem for public sector client organisations with over half (54%) rating it as highly significant compared to under a third (29%) of private sector construction organisations. Strongly linked with the affordability issues are the problems of design change orders/variations (56%) during the construction phase and establishing life cycle costs (44%) due to lack of reliable cost and benchmarking information at the planning and design development phase. As PFI is essentially a service provision on behalf of public sector client organisations, there is a concern about sustaining the service level during the operation of completed facilities. This is reflected in a rating of 48% for public sector client organisations compared to 17% for private sector construction organisations responsible for service delivery and facilities management.

The most significant issues from the perspective of the private sector construction organisations during the planning and design development phase are therefore affordability and funding gap (rated highly by 70%), inadequate client brief/requirements (60%) and poor project management (42%). At the construction phase, there are concerns about design change orders/variations (43%). However, none of the issues identified were rated highly significant during the service delivery and operation phase. These results were somehow reflected in the higher average scores for affordability/funding gap and relatively low scores (i.e. insignificant) for the construction and service delivery and operational phases.

The discourse above highlighted some of the key issues with PFI projects. There is no doubt in the UK government's agenda that PFI/PPP approach will continue to be used to provide major public infrastructure projects. However, the learning from PFI projects must be transferred to new projects to help alleviate some of the problems highlighted above. The next sections outline the findings of the survey with respect to the scope for learning and the need for the development of knowledge for capacity building.

8.6   Scope for Learning and Developing Knowledge

Based on the results from the survey (Robinson et al., 2004), over three-quarters of construction (76%) and client organisations (75.6%) agreed that there is considerable scope for learning from consortium members in the SPV/SPC. Seventy per cent of public sector client organisations and 76% of private sector construction organisations also agreed that there is scope for live capture of knowledge. There are a range of tools used for capturing and sharing knowledge in PFI/PPP projects such as skills database, content/document management systems, intranet/extranet, seminars and conferences, communities of practice, with discussion forum, post-project reviews and best practices identified as the most popular. The intranet and extranet was identified as the most important IT tool for capturing lessons learned and transferring knowledge in large organisations.

Priority areas for knowledge sharing from the public sector clients' perspective are (1) development of output specifications, (2) project management structure and monitoring mechanisms and (3) standardisation of documents. From the perspective of the private sector construction organisations, the priority areas for knowledge sharing are (1) risks, (2) design innovation/quality and (3) client brief/requirements. There are other issues such as market capacity and public perception. These areas identified by survey respondents are still considered to be major problems. There is therefore considerable scope for learning and the development of knowledge to improve PFI/PPP processes.

8.6.1   Output specifications and client requirements

A major problem in PFI/PPP projects is the development of output specification as this has a major impact on bidding process, cost and affordability of PFI/PPP projects. Pitt and Collins (2006) argued for output specifications to provide bidders with the opportunity to prioritise the service by defining the clients required level of criticality (relating to the event impacting on the asset) and functionality (relating to the assets importance). However, there are problems with defining the scope, defining the precise nature of services required and dealing with changes. Sometimes the precise definition of a high-quality service may be elusive, which allows different interpretations and can result in post-contract disputes (Akintoye et al., 2003). Subjectivity in output specifications remains a major problem as it creates different interpretations and disagreements between parties with the public sector client having one view on the performance requirement and the private sector service provider having another (4Ps, 2005). Output specifications are not always comprehensive to cover all the services required. For example, in the Darent Valley Hospital, the National Audit Office reported that the Trust have been in disagreement with the service provider regarding circumstances that were not foreseen or explicitly stated in the output specifications. The disagreement was over whether the contractor was responsible for de-icing the car park when there was an exceptionally heavy snowfall (NAO, 2005). Changes in the provision of core services provided by the public sector client can also affect the requirements set out in the output specification and can create problems. Recent studies have shown that the public sector is foregoing entitled deductions in the 'spirit of partnership' in exchange for minor contract variations (Scott and Robinson, 2008; Robinson and Scott, 2009). A key issue in PFI/PPP projects is therefore defining the scope, the need for concise definition of services required in the output specifications and clarity of the performance standards.

8.6.2   Project management structure

An appropriate project management structure is needed to ensure good governance in the delivery of PPP/PFI projects. The project management structure of PFI/PPP projects defines how power and control are cascaded throughout the organisation. For projects to be effectively implemented, it requires simple reporting structures in the public sector organisation that will allow for clear accountability and decision-making. In some cases, project management is delegated to the private sector where expertise is not available in the public sector to speed up decision-making and to reduce transaction costs. Project accountability for many public organisations is defined by the reporting structure. For example, in health sector PFI/PPP projects, the line of accountability clearly rests with the Trust's Chief Executive as the Senior Responsible Officer (NHS, 1999), and for education PFI/PPP projects, it is the local authorities and school governors. The rigorous pre-qualification and the bid evaluation process at the planning and design development phase act as powerful control mechanism to ensure that the project management structure, quality of leadership and the team selected are the most appropriate.

8.6.3   Risks

There is considerable debate on the role and valuation of risks in PPP/PFI projects in the United Kingdom. The treatment of risks in PFI models have been heavily criticised because of the subjective elements and the inconsistent application in dealing with *risk transfer* and demonstrating *value for money* (VFM). The debate and criticisms about risk allocation and VFM range from concept or definition to practical application and have come from different stakeholders including the independent National Audit Office, academics and researchers, specialist interest groups such as trade unions, public and private sector organisations. The criticisms have centred on the valuation and omission of certain risks, the process of developing the public sector comparators which one commentator described as 'malleable'. One particular issue that is now addressed in PFI/PPP projects is the problem of 'optimism bias' that is the tendency to underestimate the risks associated with cost and time overruns in public sector projects, to enable a realistic comparison with the PFI/PPP option and assessment of VFM. The UK experience shows that there is a need to have mechanisms in place for appropriate risk transfer. Incremental improvement is needed over time which will lead to more consistent approach in estimating risk and determining VFM in PFI/PPP projects. Pollock and Vickers (2002) argued that there is no standard method for identifying and measuring the values of risks to determine VFM. To facilitate a consistent approach in the estimation of risk, templates/standard documents for risk identification with a comprehensive list of transferable, negotiable and retainable risks have been developed together with a detailed quantitative methodology for estimating the cost implications of various risks.

8.6.4   Design innovation/quality

Ivory (2005) argued that there has been a general reluctance for clients in traditional procurement to encourage innovation in design as they are often unwilling to accept risks. He suggested that if consultants want to develop their market reputation, they should find 'clients that are willing to allow them to develop and "try out" new design innovation in their projects'. As Ivory (2005) puts it, 'clients should not be routinely expected to take on the risks and costs associated with innovation'. PFI projects offer the reverse scenario where significant risk and the costs associated with innovation are transferred from the client to the contractor. PFI projects therefore provide the opportunity to test the innovative behaviour of contractors but the high level of risks sometimes stifle design and technology innovation in PFI/PPP projects. Whilst there have been managerial and process innovations from the PFI/PPP contractors' perspective, introducing new designs and techniques may cost more and require more time. Design and technology innovation could generate savings in future maintenance and operational costs but there is a risk that the cost could be greater if things go wrong. Design quality has often been criticised in some PFI/PPP projects by the government's own watchdog, the Commission for Architecture and the Built Environment (CABE) noting that the expectation of design innovation from the private sector has not been forthcoming. This is possibly due to the operating environment of PFI/PPP contractors which is sometimes seen as less competitive as it is dominated by relatively few large contractors. Design development underpins all PPP/PFI projects and a number of tools have been introduced to facilitate innovation and to improve design quality such as the design development protocol and Achieving Excellence in Design Evaluation Toolkit (ADET).

8.6.5   Standardisation of documents

Standardised documentation in PFI/PPP projects is seen as a useful means of streamlining the procurement process and accelerating learning. Both the public and private sector organisations have recognised the benefit of standard documentation. This has now been developed for specific PFI sectors and areas such as contracts, pre-qualification process, output specifications for accommodation/facilities and different facilities management service areas. As the market matures, the public and private sector organisations believe that the use of standardised documents will make the PFI/PPP process become faster and cost-effective. It is increasingly recognised that 25–30 years is a long-term contractual arrangement. The public and private sector therefore need a sound contract, a good partnership/relationship underpinned by clarity in obligations which can be facilitated and enhanced through the introduction of standard documents. The need for a standard documentation in key areas of PFI/PPP projects is increasingly seen as fundamental to continuously reducing transaction costs and improving the performance of PFI/PPP projects.

8.6.6   Market capacity and public perception

Other issues identified by public sector clients and private sector construction organisations are market capacity and public perception as shown in some of the comments illustrated in Table 8.4.

Table 8.4   Selected issues relating to market capacity and public perception.

Issues	Sample commentary on key challenges
Market capacity	*'There appears to be a fairly small number of large firms that are competing for projects and as a result it is fairly difficult for new entrant to penetrate the PFI market.'* **Project Manager – local authority**
	'Lack of competition drives up prices, causes lack of innovation and cannot demonstrate value for money.' **Associate Director – large consulting organisation**
	'Maintaining a credible market – high bidding costs and uncertainties are driving participants out of the market.' **Director – consulting organisation**
	'Market overheated – too few contractors for volume of potential work.' **Managing Director – large contracting organisation**
	'Insufficient bidders to meet demand – issues for competitiveness.' **PFI Project Director – NHS Trust**
Press/public perception	*'The public are not well informed on the benefits of PFI and are only exposed to any blunders made.'* **Project Finance Analyst – large contracting organisation**
	'Poorly sold to local authorities and private sector.' **Interim Senior Education Services Manager – local authority**
	'Public opinion and bad press PFI projects receive.' **Projects Director – SPV/large contracting organisation**
	'Education of the UK public in the whole life benefits of PFI procured projects by presentation of the real debate on whole life costing. This will include education of the British press and media. Once this is resolved we will have a better chance of promoting PFI/PPP overseas.' **Director of Private Finance – large consulting organisation**

The success and failures of PPP/PFI projects has been widely reported. However, political groups tend to use it to fit their own agenda. The implication is that the benefits are not always clear to the public or taxpayer due to mixed messages and bad press. For example, the 'Times' newspaper compared the 'Evelina Children's Hospital' (a non-PFI project) with the Royal London NHS Trust development in East London. It described the former as a 'public delight' and the latter (quoting CABE) as, 'cramped', 'confusing' and seriously 'flawed' (CABE, 2005).

8.7   Developing Knowledge for Capacity Building

There are a number of problems and challenges identified and discussed in previous sections (Sections 8.2–8.6) such as bid cost, VFM, procurement timescales, barriers, output specification/clients brief, project management structure, risks, standardisation of documents, design innovation/quality, market capacity and public perception. It is important that these problems and issues are addressed through various knowledge and capacity building initiatives to improve the planning and design development, construction, and operation and service delivery phases of PFI/PPP projects.

Knowledge to continuously improve all aspects of PFI/PPP process is vital. For example, knowledge is required to assess the need and develop a business case for PFI/PPP projects, structuring, allocating and valuing risks to determine viability, affordability and bankability. Specialist skills are required to ensure that PFI/PPP projects identified are consistent with sectoral policy priorities and country strategy. PPP project development is increasingly complex, and requires considerable knowledge. Design, construction and facilities management skills are necessary for developing innovative projects and for the operation of the completed assets or facilities. Capacity for PFI/PPP development is therefore crucial to ensure appropriate projects are identified by the public sector that are attractive to the private sector in terms of the economic returns acceptable and the level of risks.

However, PFI/PPP expertise is not readily available in the public sectors, particularly in developing countries. Sometimes private sector organisations also have major capacity gaps. These problems should be addressed in the policy and strategic framework in order to embark on successful PFI/PPP programmes. Problems experienced in the early stages of the UK PPP/PFI programme resulting in a slow uptake of PPP/PFI projects was, in part, due to capacity problems. However, there was recognition of the need to accelerate learning through capacity building to speed up the implementation of projects.

A PFI/PPP framework should therefore be underpinned by appropriate capacity building strategies to identify gaps and strengthen knowledge management at local, regional and national as well as to export knowledge and experience of PFI/PPP projects overseas for gains in international trade. There are different capacity building initiatives that can be implemented to support skills development amongst individuals and strengthen organisations involved in PFI/PPP projects. A range of capacity building initiatives are shown in Table 8.5.

The World Bank Institute relied on extensive global network of strategic partnerships with public sector organisations, academia and private sector organisations to deliver its capacity building programmes (World Bank, 2006). There are specific programmes on PFI/PPP, often referred to as private participation in infrastructure (PPI) projects. The Vice President of the World Bank Institute recently argued that 'the challenge still remains with the public sector to build capacity in order to create the right environment for the private sector to invest and help discover innovation in PPP' (World Bank, 2006).

The United Kingdom is widely recognised as a leader in the implementation of PFI/PPP and many countries have learnt from it and others are expected to follow. The capacity building approach focuses on the development of explicit and tacit knowledge through (1) creation of knowledge centres and dedicated PPP units for knowledge exchange and networking, (2) development and dissemination of best practice and guidance documents to accelerate learning and improve governance, (3) traditional training and CPD events, (4) conferences, seminars and workshops, (5) staff exchange and secondment, (6) external advisers and technical assistance through for example, Partnerships UK and Department for International Development (DFID) to help other countries and (7) research and innovation. The different capacity building strategies are discussed below.

Table 8.5   Examples of capacity building initiatives.

Type of capacity building initiatives	Key features
Workshops, seminars and conference	Effective in the creation of new knowledge through tacit to tacit interaction and sharing of direct experience
Best practice documents	Useful and cost-effective in sharing explicit or codified knowledge
Video conferences	Facilitates tacit to tacit interaction with IT infrastructure playing a critical role to cut down on travel costs and associated expenses
Traditional face-to-face training courses with curriculum, course and module development	Effective in the creation of new knowledge through tacit to tacit interaction in a structured way but can be very expensive because of requirement to be at the place of delivery. Powerful in helping to increase knowledge through effective learning
Distance learning and off-campus programmes with support from Web-based and CD ROM based computer	Useful and cost-effective in sharing explicit knowledge but lack of personal or limited contact by teaching staff in some programmes is a problem. Some programmes with audio/voice recording are very effective with IT infrastructure playing a key role
Communities of practice	Informal networks of people with a common sense of purpose/community. It provides opportunity to learn from each other through tacit–tacit interaction or codification of knowledge
Staff exchange and secondment	Staff exchange and secondment are very effective approaches or knowledge transfer tools because of critical role of the direct or 'hands-on' experience
Technical assistance	Used as a mechanism to transfer knowledge from developed countries to provide support and accelerate learning in developing and transition economies
Knowledge sharing networks and e-discussion forum	Allow groups to work together and to promote information sharing through IT infrastructure
Research and development	Very useful in informing policy often using an evidence-based approach and review of practices

8.7.1   Knowledge centres

Knowledge centres play a pivotal role in knowledge exchange and networking. Their role involves documenting evidence from successful case studies and capturing lessons learned from unsuccessful PFI/PPP projects to enrich local and regional knowledge networks. This is often achieved through developing better processes and products for learning. Knowledge centres have to be created for policy advocacy, strategic planning and advisory services. For example, departmental or dedicated PPP units and support agencies are essential for the successful planning and implementation of PFI/PPP projects. In the United Kingdom, specialised PPP support agencies were created such as Public-Private Partnership Programme known as 4Ps, Partnerships UK, various units such as the central Treasury Task Force, Gateway Review teams to support the delivery of PFI/PPP projects. Such agencies continue to play a key role in policy development, implementation of best practices, providing standards, quality assurance and transparency to improve governance. Knowledge centres also promote knowledge management and exchange, through national, regional and local learning networks and knowledge partnerships, providing seminars to improve the tacit knowledge of public and private organisations involved in PPP/PFI projects. It is important to create knowledge centres to address specific issues in PPP/PFI projects. For example, excellence in design in PFI/PPP projects has been promoted by establishing programme centres to develop knowledge tools to improve design quality.

8.7.2   Dedicated PPP units

A PPP unit is defined as 'any organisation designed to promote, improve PPP, trying to attract more PPP and to ensure quality standards such as VFM, affordability and appropriate risk transfer are met' (World Bank, 2006). According to the World Bank (2007), PPP units are created to address weaknesses in the government's ability to manage programmes and to deliver key functions relating to (1) setting up PPP policy and strategy, (2) project origination and identification, (3) analysis of individual projects, (4) transaction management and (5) contract management, enforcement and monitoring. A successful PPP unit is one that contributes to the delivery of key public sector functions to achieve successful implementation.

A specialised PPP unit is one of the key components of the UK Government's policy and strategic framework to implement PFI projects. Other developed countries such as Canada, Australia, The Netherlands and Ireland introduced similar institutional structures. Specialised units have also recently evolved in middle-income and developing countries in Africa such as Egypt, Malawi, Nigeria and Tanzania (World Bank, 2007), and other countries such as Turkey and Albania embarking on PPP projects. A World Bank (2006) report noted that 'effective PPP units have tended to be attached to treasury departments (Ministries of Finance)' in parliamentary systems, reflecting its natural role in coordinating expenditure, policies and fiscal risk. For example, South Africa created a dedicated PPP unit in its Treasury Department but PPP units can be located in different government departments. The report suggests that it may take up to 3–4 years to develop appropriate and fully operational PPP units. Whilst there is some debate on the location and the particular role or function of such specialised units, it is generally useful to provide support to government departments, particularly at the initial stages when there is a need for clarification in policy, strategy and implementation as knowledge is often limited.

PPP units are often responsible for implementing and advising on PPP projects. However, the nature of the role can vary depending on the country's situation, strengths and strategy for PPP projects (World Bank, 2007). For example, in South Africa, the objective of the PPP unit is to 'filter out fiscally irresponsible PPP while creating a structure for PPP that would reassure private investors despite it being a fine filter' (World Bank, 2006, p. 5). In Australia, the objective is to improve the quality of infrastructure and to ensure that 'PPP provide for optimal risk transfer, maximise efficiency, and minimise whole life costs'. Kazakhstan recently established its dedicated PPP unit. According to EPPPC (2009), the major reason is to ensure transparency, competency and due diligence of concession projects selection process, and to accumulate advanced knowledge in PPP. The main activities of the PPP unit include economic evaluation of projects, evaluation of investment proposals, feasibility studies, bid documentation and concession agreements. However, the key responsibility for preparation of concession projects is owned by line ministries which may attract transaction advisers. It is interesting to note that according to EPPPC (2009), the Kazakhstan PPP Centre plays the 'role of external independent government "counsellor" that should ensure the balance of interests of the state, business and end-users'.

It is important that the institutional design of PPP units reflects the country's specific context such as development needs, culture, political and administrative tradition. Otherwise, it may not be successful in providing advice and in the implementation of projects. The World Bank (2007) argued that the 'one-size-fit-all' approach is not appropriate in the design of dedicated PPP units. There is also a recognition that developed countries and their advisers should not impose assumptions relating to the development of PPP units without understanding the specific context and challenges in developing countries. For example, resources and skills are often a major problem as resource markets are not always efficient which can undermine the cost of infrastructure development.

8.7.3   Best practice and guidance documents

Government and support agencies are responsible for supporting PFI/PPP policy, using evidence gathered through monitoring and evaluation of PFI/PPP projects in different sectors to accelerate learning and the development of new knowledge to improve PPP/PFI practices. Developing best practice and standard documents is important to address a range of issues/themes relating to the planning and development, service delivery and operational phases of PFI/PPP projects. For example, there are standard contract documents, templates for developing output specifications to reflect accommodation standards and FM services, pre-qualifying bidders and selecting consultants for advisory services. Best practice and guidance documents are designed to improve governance in the delivery of PFI/PPP projects. In the United Kingdom, for example, a key issue central to justifying PPP/PFI projects is the VFM test. There has been a considerable debate on how to operationalise the VFM criteria in justifying PFI/PPP projects. Problems in this area were addressed through the development of a number of standard guidance documents and technical notes to explain how VFM should be calculated. VFM is assessed during the planning process from the outline business case to the full business case in accordance with the guidance and procedures set out through a revised document (HM Treasury, 2006) to ensure a consistent interpretation and application of the VFM criteria. Other best practice initiatives include the design quality guide 'Achieving

Excellence in Design Evaluation Toolkit (ADET)' and the development of a 'compendium' of exemplar designs for new primary and secondary schools for the then Department of Education and Skills (DfES) to produce model schemes for different types of site.

8.7.4   Traditional training and CPD events

It is essential that public sector and private sector staff involved in PPP/PFI projects participate in various training events which can be delivered face-to-face or through distance learning to improve their skills as part of staff appraisal. Some training programmes are more structured and involve identifying PFI/PPP skills set and competencies required to complete the course. An example of this in the United Kingdom is the specialist diploma programme proposed to address the specific training needs of PPP/PFI project directors. As part of this training initiative, 4Ps partnered with a UK University and Constructing Excellence to design and deliver the PFI/PPP project directors training programme (4Ps, 2008). The programme focuses on developing leadership skills, technical expertise in PFI/PPP and other complex projects in local government. 4Ps also has other stand-alone training and development modules to address the specific needs of directors, managers and others involved in PFI/PPP projects at various levels – strategic, project, specialist and operational levels. Other training programmes relevant to PFI/PPP projects are often delivered as part of a module or unit in established postgraduate courses in various universities.

Continuous professional development (CPD) events on PPP/PFI are regularly organised by professional bodies. This includes the Chartered Institute of Building (CIOB), Royal Institution of Chartered Surveyors (RICS), Institute of Civil Engineers (ICE), Royal Institute of British Architects (RIBA), Institute of Chartered Accountants in England and Wales (ICAEW) and other international professional bodies such as the International Road Federation (IRF). For example, the IRF recently organised a conference on 'Road Safety and PPP' in Egypt.

International programmes are also available from private organisations such as courses offered by the Institute of Public-Private Partnership (IP3) in the United States. IP3 courses offer a wide range of time-specific online/distance learning courses and short-term face-to-face courses from basic courses dealing with fundamental principles of PPP, developing PPP competencies, to advanced modules such as the 6-week intensive course on 'Advanced Project Finance and Financial Modelling'. Under this training programme, credits can be accumulated to become a Certified Public-Private Partnership specialist after a period of training (IP3, 2009). The IP3 course is an example of innovative 'online' training programme that uses the Internet and an IT platform with various learning tools to deliver modules and course materials which are available '24/7', and to facilitate interaction with course participants and the course facilitator. A key advantage is providing training whilst on the job and significant reduction in the expenses associated with traditional training programme.

8.7.5   Knowledge sharing networks and communities of practice

Training can also be acquired through knowledge sharing networks such as UNDP Global Learning Network (GLN). The UNDP PPP for urban environment (PPPUE) learning network facilitates the exchange of knowledge between PPP practitioners which can help to implement public-private partnerships successfully. There are a number of features in the GLN that facilitates exchange of lessons learnt, ideas and the development of knowledge through participating in the 'Experience Exchange' (UNDP, 2008). The GLN provides opportunities for PPP colleagues to interact virtually or in person. A number of knowledge management tools are used to support the network or communities of practice such as the GLN News e-mail forum to receive PPP Newsletters, the Consultant Roster, or through participation in upcoming events shown in the PPP Events Calendar.

There is also evidence of other knowledge sharing networks gradually evolving. For example, a team of PPP specialists which held their first meeting in February 2008 was recently formed by the United Nations Economic Commission in Europe (UNECE) to create a mechanism for the effective exchange on knowledge, taking stock of recent developments in PPP agreeing on a work programme for implementation of its PPP activities (UNECE, 2008). The three main activities identified by UNECE are to disseminate best practices and to raise awareness of the benefits of PPP, to provide training of public and private sector officials and policy and project advice such as assisting governments in establishing PPP units, developing policies and strategies, coordination between agencies involved in PPP. The International Council for Research and Innovation which acts as a global network for international exchange and cooperation in research and innovation in the built environment has also launched CIB TG 72 on PPP in August 2008. The task group was formed in recognition of the growing interest of PPP in both developed and developing countries and the need for an independent international forum 'to exchange and help synthesise research on underpinning and emerging issues' (CIB, 2008).

Private sector organisations involved in PPP projects have communities of practice dealing with different issues and interest across a range of sectors. Public sector client organisations also have similar groups or communities of practices such as the Education Building and Development Officers group dealing with a variety of procurement topics relating to school building.

8.7.6   Conferences, seminars and workshops

Training events can sometimes take the form of conferences, workshops and seminars drawing on the strengths of different organisations in delivering an event. For example, the European PPP Centre in cooperation with IRF recently organised a highly successful conference on 'Promoting PPP Investments in Central Eastern Europe (CEE)' in Budapest, Hungary (EPPPC, 2007). Also, a special workshop focusing on a particular motorway project in

Hungary was recently organised by EPPPC in cooperation with the Czech PPP Association. These events often drawing on the experience or tacit knowledge of high-level project participants from the public sector and private sector based on case studies of actual projects provide a powerful and interactive platform for tacit–tacit knowledge exchange, learning and lively discussions. There is the benefit of a detailed understanding of the context of how knowledge was developed in specific PPP projects.

There are also other capacity development programmes offered by multilateral development agencies such as the World Bank, African Development Bank, Asian Development Bank and Inter-American Development Bank. For example, the World Bank Institute, the knowledge management and capacity building arm of the World Bank, organised the first international gathering of PPP units in Washington in 2006 as a forum to discuss and share information and experiences of key issues in PPP projects, successes and failures and to promote a global network of PPP units around the world (World Bank, 2006). The event attracted 40 participants (excluding staff from World Bank and representatives of sister organisations) from 23 different PPP units from developing and developed countries reflecting different levels of PPP maturity and scale from advanced, medium and small PPP programmes. Countries represented included the United Kingdom, Australia, Japan, Korea, Uganda, Pakistan, Nigeria, Mexico and Peru amongst others.

Specialist 'practitioner-oriented' and academic conferences on PFI/PPP are also available. These events are often delivered by various organisations such as public, private, universities and professional organisations. For example, SMi have had an established portfolio of PPP/PFI events running since the mid-1990s. They organise conferences, workshops and other events that are research driven and targeting senior level professionals to learn and acquire knowledge about the latest techniques and developments in PPP/PFI projects. SMi's PFI/PPP conferences cover topics ranging from financial modelling and accounting through to identifying new projects and markets in countries such as Ireland, The Netherlands, Portugal, Italy and Spain (SMi, 2009).

8.7.7   Staff exchange and secondment

Other training methods or knowledge transfer tools can be used such as on-the-job training, staff exchange and secondment, and twinning initiatives. HM Treasury (2006) identified ways in which the government could improve the PFI procurement process including the development of a secondment model within the public sector so that public servants with tacit knowledge or experience of complex procurements can be retained and deployed on projects across the public sector to facilitate the transfer of knowledge. Staff exchange and secondment are very effective approaches or knowledge transfer tools because of the critical role of the direct experience to facilitate the development of tacit knowledge in other groups. The contribution to organisational knowledge in delivering actual projects is immense as mistakes can be costly to the organisations.

8.7.8   External advisers and technical assistance

Public sector organisations tend to rely heavily on external advisers from the private sector to provide a range of legal, technical and financial advice from planning, construction to the operational phases of PFI/PPP projects due to lack of in-house knowledge. In the United Kingdom, as in many other developed countries, such advice is often crucial to identify appropriate PFI/PPP projects and to facilitate implementation.

External advisers therefore play a key role in the development of tacit knowledge and capacity building in the public sector organisations.

In the developing and transition economies, the role of multilateral development banks and agencies is crucial in providing technical assistance and facilitating the dissemination of best practices and guidance documents. Technical assistance is used as a mechanism to transfer knowledge and accelerate learning in developing and transition economies. For example, the limitation of infrastructure policy and development strategies prompted the establishment of the Public-Private Infrastructure Advisory Facility (PPIAF) to help improve infrastructure delivery. The PPIAF was launched in 1999 as a joint initiative of the governments of Japan and United Kingdom in association with the World Bank. It is a multi-donor technical assistance facility governed by a council consisting of participating donors, which includes bilateral agencies from developed countries, multilateral development agencies and international financial institutions. The council is supported by an independent Technical Advisory Panel, consisting of international experts, knowledgeable in various aspects of private sector involvement in infrastructure. The objectives of PPIAF are fulfilled through two main channels:

- Providing technical assistance to develop infrastructure policies and strategies to fully engage the private sector through PPP
- Disseminating best practices of private sector involvement in the delivery of infrastructure services

PPIAF provides technical support to facilitate private sector involvement from financing, rehabilitation, operation and management of services. They fund a range of activities in developing and the transition economies of Africa, Asia, Eastern Europe, Middle East, Latin America and Caribbean focusing on certain types of technical infrastructure – roads, ports, airports, railways, electricity, solid waste, water and sewerage, telecommunications and gas transmission. They also fund studies to help countries develop a policy and strategic framework and action plan for governments and donors to determine reform and investment priorities. For example, a recent study titled 'Public-Private Partnership Units: Lessons for Their Design and Use in Infrastructure' aimed at the development of effective and dedicated PPP units was carried out jointly with the World Bank (2007).

8.7.9   Research and innovation capacity

The nature of the problems and challenges involved in the planning and development, construction, service delivery and operational phases of PFI/PPP projects means that there is a need for research to inform policy and practice. There are significant opportunities for innovation from research projects but design and construction firms often invest relatively little in research and development (R&D) compared to other sectors. Reichstein *et al.* (2005, p. 642) argued that 'construction firms have become inherently risk averse and many construction firms do not need to innovate to remain successful'. This could, in part, explain why there has been a low level of design innovation in PFI/PPP projects, given that many of these large firms involved in PFI/PPP projects are risk averse. Lack of qualified personnel has been identified as a major factor affecting the level of innovation in design and construction firms. It has therefore been suggested that there is a need for 'innovation champions' to take the role of integrator between clients, industry and contractors or knowledge broker to facilitate the uptake of new knowledge from research activities carried out internally and from external sources (Winch, 1998).

Universities have continued to play a key role in facilitating innovation through research, development and consulting. Research projects have been funded by the UK research councils and other agencies including the private sector to investigate and improve specific aspects of PFI/PPP projects such as bidding, risk management, life cycle costs, VFM, accountability and innovation. As a result, incremental improvements have been made in specific areas of the PPP/PFI process. Whilst there have been some progress in process innovations, the level of design innovation in PPP/PFI projects is generally mixed and at times disappointing from the perspective of certain stakeholders such as the government's watchdog, CABE. There is therefore a need to develop R&D and innovation capacity to improve the performance of PFI/PPP projects. This is necessary to address significant challenges in PPP/PFI projects to increase project performance, efficiency, sustainability, project management, knowledge transfer and to reduce the high bidding and transaction costs associated with PPP/PFI projects. Research will continue to play an important part in the innovation and development of PPP/PFI processes, in particular in the development of design and construction technologies to improve service delivery. Due to the increasing popularity of PPP, greater attention is now focused on improving whole life cycle cost and performance, together with associated capital and operating costs.

8.8   Concluding Remarks

Capacity building is crucial for the development and acquisition of the managerial, financial, legal and technical skills to successfully implement PPP/PFI projects. Lack of skilled resources slows development and implementation programmes with significant negative impact on the speed, efficiency and cost of delivering PPP/PFI projects. A range of capacity building initiatives are essential to develop relevant skills, tacit and explicit knowledge. Knowledge centres and support agencies such as Private Finance Unit, Office of Government Commerce (OGC), Partnerships UK and 4Ps have been established in the United Kingdom to play a catalytic role in the implementation of PPP/PFI projects by developing capacity for public sector clients on all aspects of PPP including the development of output specification, bid documents, evaluation of risks and VFM and to facilitate the dissemination of best practices.

The next chapter discusses the application of a knowledge transfer framework to accelerate learning and capacity building in organisations. The framework is useful in identifying appropriate knowledge management tools and techniques to capture the type of knowledge required in an organisation, whether tacit or explicit, whether it is available internally or externally, resides with individuals or specific groups to implement appropriate learning and capacity building initiatives.

References

4Ps (2005) *4Ps Review of Operational PFI and PPP Projects.* Available online at www.4ps.gov.uk.

4Ps (2008) *Training Programme for Project Directors.* Available online at www.4ps.gov.uk.

Akintoye, A., Hardcastle, C., Beck, M., Chinyio, E., and Asenova, D. (2003) Achieving best value in private finance initiative project procurement. *Construction Management and Economics* 21, 461–470.

Ball, R., Heafey, M., and King, D. (2000) Private finance initiative – a good deal for the public purse or a drain on future generations? *Policy and Politics* 29, 95–108.

CABE (2005) *News, Final Views on the Royal London Hospitals Proposals.* Available online at http://www.cabe.org.uk (Accessed 18 September 2006).

CIB (2008) *Workshop on Revamping PPP's: From 'Revisiting and Rethinking' to 'Revamping and Revitalising' PPP's.* CIB News Article on Task Group TG72 – Public Private Partnership, December 2008.

Davies, J. (2006) *Risk Transfer in Private Finance Initiatives (PFIs) – An Economic Analysis.* DTI, Industry Economics and Statistics Directorate (IES) Working Paper, February.

de Lemos, T., Almeida, L., Betts, M., and Eaton, D. (2003) An examination on the sustainable competitive advantage of private finance initiative projects. *Construction Innovation* 3(4), 249–259.

European Public Private Partnership Center (EPPPC) (2007) *Promoting PPP Investments in CEE.* Budapest, Hungary. Accessed 29–30 October 2007.

European Public Private Partnership Centre (2009) *About Us.* Available online at http://www.epppc.hu/about_us (Accessed 24 February 2009).

Ezulike, E.L., Perry, J.G., and Hawwash, K. (1997) The barriers into the PFI market. *Engineering Construction and Architectural Management* 4(3), 179–193.

HM Treasury (2003) *PFI: Meeting the Investment Challenge.* The Stationery Office, London.

HM Treasury (2006) *PFI: Strengthening Long-Term Partnerships.* HM Treasury, London.

Institute of Public Private Partnership (IP3) (2009) *On-Line Training Programme.* Available online at www.ip3.org.

Ivory, C. (2005) The cult of customer responsiveness: is design innovation the price of a client-focussed construction industry? *Construction Management and Economics* 23, 861–870.

NAO (National Audit Office) (2001) *Managing the Relationship to Secure a Successful Partnership in PFI Projects.* The Stationery Office, London.

NAO (National Audit Office) (2005) Darent valley hospital: the PFI contract in action. *Report of Comptroller and Auditor General, HC 209, Session 2004–2005.* The Stationery Office, London.

NHS (National Health Service) (1999) *Public Private Partnerships in the National Health Service: The Private Financial Service: Good Practice.* HMSO, United Kingdom Official Publications Database, London.

Pitt, M., and Collins, N. (2006) The private finance initiative and value for money. *Journal of Property Investment and Finance* 24(4), 363–373.

Pollock, A., and Vickers, V. (2002) Private finance and value for money in NHS hospitals: a policy in search of a rationale? *British Medical Journal* 324, 1205–1208.

Reichstein, T., Salter, A.J., and Gann, D.M. (2005) Last among equals: a comparison of innovation in construction, services and manufacturing in the UK. *Construction Management and Economics* 23, 631–644.

Robinson, H.S., Carrillo, P.M., Anumba, C.J., and Bouchlaghem, N.M. (2004) *Investigating Current Practices, Participation and Opportunities in Private Finance Initiative.* Loughborough University, Leicestershire.

Robinson, H.S., and Scott, J. (2009) Service delivery and performance monitoring in PFI/PPP projects. *Construction Management and Economics* 27(2), 181–197.

Scott, J., and Robinson, H. (2008) Payment mechanism in operational PFI projects. In: Akintoye, A., and Beck, M. (eds). *Policy, Management and Finance for Public Private Partnerships.* Wiley-Blackwell Publishing, Oxford, pp. 414–435.

SMi (2009) *PFI and PPP Training.* Available online at http://www.smi-online.co.uk/training/default.asp?pb_sect=3 (Accessed 24 February 2009).

UNDP (2008) *PPP for Urban Environment (PPPUE).* Available online at www.undp.org.

United Nations Economic Commission of Europe (UNECE) (2008) *Report of a Team of Specialists on Public Private Partnerships on its First Session, ECE/CECI/PPP/2008/2.* Economic and Social Council, Geneva, 28–29 February.

Winch, G. (1998) Zephyrs of creative destruction: understanding the management of innovation in construction. *Building Research and Information* 26(4), 268–279.

World Bank (2006) Proceedings from PPPI Days 2006, June 8–9, Organised by the World Bank Institute. The World Bank, Washington, DC.

World Bank (2007) *Public Private Partnership Units: Lessons for Their Design and Use in Infrastructure, Sustainable Development in East Asia and Pacific.* The World Bank, Washington, DC.

9

The Knowledge Transfer Framework

9.1   Introduction

By March 2008, the UK Government had signed over 628 PFI/PPP projects totalling more than £58 billion. Given this level of public spending, there is pressure on the government to ensure its funding is spent wisely. Thus, the government's spending watchdog – the National Audit Office (NAO) – has produced over 50 reports on PFI/PPP projects in an effort to improve the delivery process. In addition, other bodies such as Partnerships UK provide advice on PFI procurement; they have also produced several guidance notes on contractual documents and host a database of PFI/PPP projects on a regional basis. Another body, 4Ps (Public-Private Partnership Programmes), works with local authorities to improve the PFI/PP delivery process.

Despite these initiatives, there is still scope for improvement. For example, NAO (2007) stated that, '[S]ystematic ways to ensure useful lessons are shared were not always exploited.' It was further recommended that, 'Departments should identify lessons from recently closed PFI/PPP projects of relevance to subsequent projects.' The same report later highlighted that, '[T]here should be a more structured process of learning and sharing lessons across sectors.' Thus, a key issue with PFI/PPP projects is the knowledge transfer that occurs both within and between projects.

This chapter discusses the need for knowledge transfer framework to enable a structured approach to learning. This was part of a study undertaken to investigate the level of participation on PFI, to explore opportunities for construction organisations and to understand the mechanisms for enhancing knowledge transfer. The findings with respect to the key issues and knowledge management challenges have been discussed in the previous chapter. The chapter starts with a discussion of the key issues relating to knowledge transfer. The conceptual model is presented and the different components of the knowledge transfer framework to facilitate learning and capacity building are examined. The key stages involved in the development of a knowledge transfer framework and specific issues are identified. The framework helps to identify knowledge gaps and the scope for learning to expand business opportunities. The role of knowledge management tools and the importance of monitoring learning to assess the effectiveness of knowledge transfer activities are examined. The framework concludes with the need for an action plan to address barriers to knowledge transfer and to implement appropriate learning and capacity building activities to facilitate continuous improvement. The evaluation of the framework is also discussed in the chapter.

9.2   Knowledge Transfer Issues

Knowledge transfer is an area of increasing interest to many organisations, particularly those involved in PFI/PPP projects. Argote *et al.* (2000) provide a summary of the various mechanisms available. These include personnel movement, training, communication, observation, technology transfer, alliances, and so on. A number of authors have also proposed models or frameworks to enhance knowledge transfer (Argote and Ingram, 2000; Szulanski, 2000; von Krogh *et al.*, 2001; Goh, 2002). However, these have not yet filtered into the construction sector and hence have not yet been exploited. One of the reasons may be that these frameworks are often at a conceptual level (Argote and Ingram, 2000) and highlight factors to consider, rather than practical actions for a firm or organisations to address. For example, Goh (2002) highlights factors such as leadership, problem-solving/seeking behaviours, support structures, absorptive and retentive capacity, and types of knowledge.

PFI/PPP projects are increasingly popular in the United Kingdom and other countries but the need for continuous improvement is recognised as crucial. There were major challenges associated with the early PFI/PPP projects in the United Kingdom and the need to transfer lessons learned for improvement was considered critical for future projects. This may somewhat explain why an Ernst and Young report (2002) argued that, '[I]t is perhaps a good time to reflect on how PFI has developed and why it has turned out to be more challenging than the original enthusiasts thought.' The report indicated that there are still concerns over the level of knowledge sharing. The Audit Commission (2003) highlighted the need for the early lessons learned in PFI to be 'recycled effectively during future investment' to improve performance. HM Treasury (2004) also stressed the importance of information sharing for the better performance of PFI projects. However, there is a need to determine knowledge transfer needs or knowledge gap based on opportunities in the PFI/PPP market and the types of knowledge required. It is also important to understand the mechanisms for knowledge transfer and some of the challenges involved and barriers associated with transferring knowledge in order to develop an effective knowledge transfer framework.

9.2.1   Knowledge transfer needs

All projects require knowledge transfer but the need is even more critical for organisations involved or interested in PFI/PPP projects to enter into particular areas of PFI/PPP market or expand their PFI/PPP work. The main reason for this is that it is a relatively new form of procurement, all parties are new to the process and there is a shortage of expertise in this area. PFI is a costly commitment; hence, any mistakes made because of lack of current knowledge can be critical for the length of the service period of the contract. In PFI, all parties are learning and the PFI process is continuously evolving as seen by the need for bodies such as 4Ps to provide support for local authorities.

Organisations involved in PFI/PPP alliances such as the special purpose company/vehicle (SPC/SPV), client, consultants, contractors and facilities management organisations could benefit significantly from knowledge transfer. Studies show that a significant proportion of construction organisations recognise the benefits of knowledge transfer such as reducing rework, avoiding re-inventing the wheel, improved utilisation of tacit knowledge and best practices to facilitate continuous improvement and innovation (Robinson et al., 2001; Carrillo et al., 2004). Other organisations such as clients in the health sector, transport, education and other sectors with major in PFI/PPP projects and programmes also have a similar need. In fact, the need for knowledge transfer tends to be greater in client organisations due to their limited experience in PFI/PPP projects (Robinson et al., 2004). Knowledge transfer could also be an effective mechanism for mitigating risks, a key issue in an increasingly complex PFI/PPP environment. However, the implementation of a knowledge strategy is still underdeveloped in client, design and construction organisations. A key challenge is, therefore, to address what knowledge needs to be transferred and how best to do so.

9.2.2   Types of knowledge to transfer

The Robinson et al. (2004) study highlighted the planning and development phase, sometimes called the procurement phase, as the most problematic area requiring both knowledge creation and knowledge sharing/transfer. The findings from the case studies discussed in Chapter 7 and the results of the questionnaire survey presented in Chapter 8 clearly demonstrate the need to develop knowledge in key areas such as output specification/client brief, staff transfer from public to private sector, contractual issues, town planning and land issues, construction and decanting issues, existing asset and facilities data, budgeting/costing to deal with affordability, financial modelling and payment mechanism. For example, construction companies point out that they have little data for costing the maintenance of a facility over a period of 20–30 years. This knowledge of life cycle cost estimating is critical if sensible estimates are to be made which do not exacerbate the affordability/funding gap problem in PFI/PPP projects. In housing refurbishment PFI, there are major problems with stock condition data crucial for estimating a realistic cost of the PFI projects. Adams (2005) noted that bids for PFI credits by local authorities were based on scant stock condition surveys. The business plans developed to apply for government PFI credits, therefore, create affordability problems as they often underestimate the true costs of the capital works and recurrent maintenance expenditure. This is because surveys have been based on a 'cloning technique' with a relatively small cross-section of

properties assessed to determine the scope of work. Adams (2005) suggested
that all parties should invest in good quality surveys to improve knowledge
on existing asset and facilities, an issue which remains a major problem in
refurbishment PFI/PPP projects. There are also other major problems such
as developing output specification and estimating the value of risk transfer
which is used as the basis for determining and comparing the public sector
comparator (PSC) or private sector PFI bid. Payment or unitary in PFI/PPP
projects is normally made up of several elements including a service component of the charge through benchmarking and market testing. However, data
on benchmarking are sometimes difficult to get. The knowledge required can
be explicit (codified) which is easy to transfer or tacit which is more difficult to transfer without understanding the context. In addition, because
some PFI projects consist of the construction of multiple structures/facilities
(e.g. a number of schools for a local authority), there is a need to transfer
lessons from one project into future building projects. Thus, there is a need
to not only transfer knowledge throughout the life cycle of the project but
also transfer the lessons learned from one PFI project to another that may be
happening simultaneously or with a (limited) time lag.

9.2.3   Mechanisms for knowledge transfer

There are a number of mechanisms for sharing or transferring knowledge.
These tend to fall into two main categories – tools and techniques (Al-
Ghassani, 2003). Tools rely on the use of information technology (IT) to
share typically explicit knowledge, that which is easy to document and store.
Examples are databases and project extranets used in collaborative construction projects. Techniques adopt a more human-centric approach for transferring mainly tacit knowledge, that which is based on expertise and intuition
and is difficult to transfer. Typical examples of techniques recommended
for PFI/PPP projects to transfer knowledge are communities of practice and
post-project reviews. An appropriate mechanism is therefore required depending on the type of knowledge to transfer and the other characteristics
such as whether the knowledge is available internally or externally, resides
with particular individuals or groups. Understanding the characteristics of
knowledge helps to determine which tools or techniques are suitable for
knowledge transfer. But different tools will be required in different phases or
stages in PFI/PPP projects. Case studies with engineering design firms show
that about 80% of knowledge used during concept design is tacit compared
to about 20% of explicit knowledge; the reverse is true at the detailed design
stage with 20% tacit and 80% explicit (Al-Ghassani, 2003). Due to the complexity of PFI/PPP projects and as a relatively new form of procurement, there
is considerable amount of tacit knowledge required. Knowledge developed
over time in PFI/PPP projects becomes codified (explicit knowledge) in the
form of best practice and guidance documents available externally. Explicit
knowledge is constantly evolving and disseminated by major government
and support agencies such as the Treasury, Office of Government Commerce
(OGC), 4Ps and Partnerships UK. There is now increasing amount of advice

from bodies such as the Construction Excellence (CE) and the Construction Industry Research and Information Association (CIRIA) on how knowledge can be shared and the types of techniques and tools available. However, Brooking (1996) pointed out that only 20% of an organisation's knowledge is actually used whilst Newell *et al.* (2002) highlighted the need for organisations to have a supportive organisational culture and trust to encourage knowledge sharing. The techniques and tools facilitate knowledge transfer activities. However, it is often difficult to select the most appropriate tools for a particular organisation due to the significant growth of sophisticated technologies available in the market place. The challenge is identifying which mechanism best suits the organisational context.

9.2.4   Knowledge transfer problems

Knowledge sharing networks in alliances such as those created to execute PFI/PPP projects raise complex issues such as confidentiality, reliability, copyright, the dissemination of a firm's unique stock of knowledge outside its boundaries, and the trade-off between cooperation and competition or what is referred to as 'co-opetition' (Levy *et al.*, 2001). The ability to learn is also crucial to effective knowledge transfer, and an organisation's absorptive capacity to manage new knowledge depends on prior knowledge and technical capability (Gann, 2001). Learning starts at an individual level, building individual technical capabilities to become a knowledge organisation. Knowledge transfer can facilitate the creation of learning networks that are spread throughout organisations (McAdam and McCreedy, 1999) that are necessary for the improvement of skills and competencies to support the delivery of PFI projects. Thus, PFI/PPP projects designed to operate beyond organisational boundaries should provide a stimulus for knowledge sharing and innovation. Bresnen and Marshall (2000) noted that some organisations are willing to share technical know-how with partners, whilst others may jealously guard such proprietary information. Crawley and Karim (1995) used the term 'permeable boundaries' to describe the flow of appropriate resources from one organisation to another and the restriction of leakage of sensitive and confidential information. There are also other issues that are critical, such as an organisation's readiness to transfer knowledge. Organisational readiness relates to both hard (e.g. resource requirements, IT infrastructure, hard performance measures) and soft (e.g. organisational culture, incentive structure, trust, soft performance measures) issues necessary for knowledge transfer to be successfully implemented. Relying on 'goodwill knowledge philanthropy' that knowledge transfer can take place without a proactive approach involving creating knowledge sharing networks, enhancing learning capacity and other support mechanisms have been shown to be ineffective. The long-term commitment in PFI/PPP projects provides an opportunity for construction organisations to take a stake in continuously improving the PFI project delivery processes and the constructed facilities. The relatively small number of construction organisations which include consultants and contractors involved in PFI, the repetitive nature of PFI in specific sectors, alliances created and long-term relationships with clients and other stakeholders can provide a stimulus for learning, knowledge transfer and innovation.

9.3   The Knowledge Transfer Conceptual Model

Any knowledge transfer framework should therefore consider the needs for an organisation in terms of business opportunities, the types of knowledge required, the mechanisms to transfer knowledge to those who need it and how to address the specific problems and barriers associated with knowledge transfer activities. A collaborative research project was undertaken by Loughborough University, and a group of construction/PFI companies and clients aimed at delivering to clients and construction organisations a toolkit for improving knowledge transfer on PFI projects. First, a conceptual model was developed based on findings from survey and case studies to address specific industry's needs. The conceptual model was evaluated in a workshop with seven of the project's industry collaborators. Following the feedback from industry evaluation workshops, the conceptual model was subsequently developed into what was called a 'knowledge transfer framework' to facilitate learning and capacity building.

The conceptual model consists of three stages as shown in Figure 9.1. Stage 1 provides a structure to review current PFI practices and identifies the scope for learning in order to improve PFI participation and explore further opportunities in PFI. Stage 2 investigates knowledge transfer problems in terms of the knowledge characteristics, knowledge transfer mechanisms and barriers to knowledge transfer. Stage 3 aims to develop action plan to develop a learning culture to support continuous improvement in PFI projects. The conceptual model was evaluated by the project's industry collaborators and subsequently developed into a knowledge transfer framework.

The knowledge transfer framework should be used collaboratively and involve PFI/PPP staff, business development managers and knowledge managers. Two of the three stages require supplementary documentation

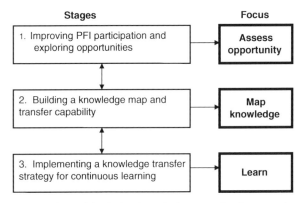

Figure 9.1   Conceptual model of the knowledge transfer framework. From Carrillo *et al.* (2006); reprinted by permission of the publisher, Taylor & Francis Ltd (http://www.informaworld.com).

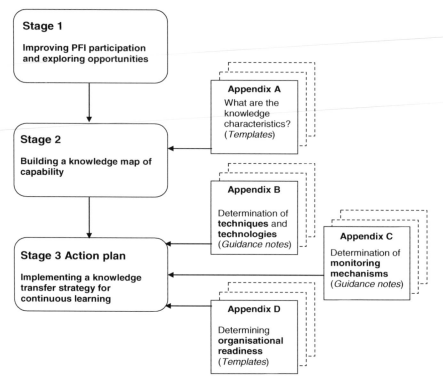

Figure 9.2 The knowledge transfer framework flow chart. From Carrillo *et al.* (2006); reprinted by permission of the publisher, Taylor & Francis Ltd (http://www. informaworld.com).

or detailed guidance notes that were provided in the form of supporting appendices as shown in Figure 9.2.

The knowledge transfer framework was made more user-friendly by providing a colour-coded flow chart. Figure 9.2 shows the flow chart indicating how the three stages and their supporting guidance notes and appendices fit together.

9.4   Improving PFI/PPP Participation and Exploring Opportunities (Stage 1)

The aim of this stage is to provide a structure to review current practices and identify the scope for learning to improve PFI/PPP participation and explore further opportunities. The outcome of this stage is a form that identifies key issues in a PFI/PPP stage that need to be addressed regarding knowledge transfer. It also identifies current knowledge transfer practices, how these may be improved and the scope for learning and knowledge transfer associated with respect to other PFI stages and other PFI sectors. A worked example of the Stage 1 form is shown in Table 9.1.

The form is used to specify the problem in its proper context, by relating it to the phase and particular stage of PFI it affects. The template also enables the identification of the specific issue; in this case, the issue relates to 'afford-

Table 9.1 Sample stage 1 form for exploring scope for improvement and opportunities.

Stage	Tasks	
1.1	PFI stage to consider	Outline business cases (OBC)
1.2	Description of issue	Affordability based on quality of output specification
1.3	Identify the PFI sector that the issue relates to	Education and health
1.4	What are the current practices with respect to the issue?	Limited funding from central government Technical standards and specifications dictate output Balance of funding, therefore; needs to come from other sources Incorrect advice from consultants who are not aware of recent standards
1.5	Identify how current practices can be improved	Schemes may have to be re-scoped, for example two not three schools Review facility management standards Increase council tax
1.6	Identify the scope for learning/knowledge transfer associated with the issue	Benchmarking using regional data and type of site. High scope for learning.

From Carrillo *et al.* (2006); reprinted by permission of the publisher, Taylor & Francis Ltd (http://www.informaworld.com).

ability' linked to the output specification. The particular sector it relates to (in this example, health and education) so that the problem can be understood within the context of health and education. The current understanding of the factors affecting the issue of affordability is also identified such as limited funding and standards relating to the output specification. Output specifications that require services far in excess of what is intended can lead to affordability problems. The need to look for other sources of funding (e.g. sale of land, third party revenue from PFI/PPP scheme) and incorrect advice from consultants is also identified as factors relating to affordability.

The template also enables the identification of current practices and the knowledge required to address the problems of affordability, which includes reducing the cope of schemes/output specification, the facilities management service standards to reduce cost and improve affordability. Benchmarking knowledge is identified as one example of knowledge transfer activity that could lead to more accurate and reliable costing of facilities management services which could reduce or eliminate the problem of affordability associated with PFI/PPP bids. Other examples of knowledge transfer activity could be identified such as reducing the scope through the development or more appropriate output specification. The appointment of better advisers with more up-to-date technical knowledge and identifying other sources of income or third party revenue from PFI/PPP schemes will reduce or eliminate problems of affordability. The process is repeated for all other issues affecting the different phases (from planning and development, construction, service delivery and operational) and specific stages of PFI/PPP projects to form the basis of identifying future opportunities. However, this will require the development of a knowledge map for the organisation.

9.5   Building a Knowledge Map and Transfer Capability (Stage 2)

The aim of this stage is to investigate knowledge transfer issues in terms of what knowledge needs to be transferred: (1) its characteristics, (2) transfer mechanisms and (3) barriers to knowledge transfer. The knowledge characteristics are determined using a supplementary guidance note or appendix that asks users to determine the characteristics of the knowledge to be transferred based on classifications with a sliding scale as shown in Figure 9.3.

First, the knowledge characteristics template helps to determine the knowledge transformation trajectory or path from the current state of the knowledge to the future knowledge requirements in terms of whether the knowledge is available as individual/shared knowledge, externally/internally or tacit/explicit knowledge. Second, it is important to identify the transfer mechanism, knowledge tools (IT systems) or techniques (non-IT systems) to support the transformation process from the current state to the future state required using the knowledge characteristics template. Third, the barriers to facilitate knowledge transfer activity should also be identified such as issues relating to confidentiality, reliability, copyright and availability of knowledge. Individual, team and organisational barriers should also be addressed.

The outcome of this stage is a form that identifies the type of knowledge that should be transferred, the characteristics of this knowledge from current practices to future requirements, transfer mechanisms and barriers to transferring knowledge from current to future state to facilitate improvement in dealing with the particular issue (e.g. benchmarking of hard and soft facilities management costs) and to reduce the problems of affordability in PFI projects. A worked example of the Stage 2 form is shown in Table 9.2.

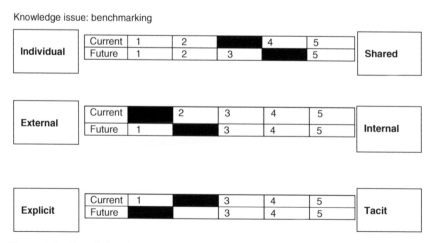

Knowledge issue: benchmarking

Figure 9.3   Knowledge characteristics template. From Carrillo *et al.* (2006); reprinted by permission of the publisher, Taylor & Francis Ltd (http://www.informaworld.com).

Table 9.2 Sample stage 2 form for building knowledge map and capability.

Stage	Tasks	
2.1	Type of knowledge required	Benchmarking
2.2	State the *current* characteristics of the knowledge for each type listed in 2.1	Benchmarking knowledge is mainly (i) individual (ii) external (iii) tacit
2.3	What are the current mechanisms/ways for sharing this knowledge	Benchmarking knowledge: IT systems and Black book
2.4	Identify the barriers currently associated with existing mechanisms	Quality of information. It is not reliable and confidentiality issues exist
2.5	Identify the *future* knowledge characteristics	Benchmarking knowledge should move from (i) individual to shared (ii) external to internal (iii) tacit to explicit
2.6	Identify barriers relevant to moving from the existing to future characteristics	Availability of resources Leads to single-point expertise Cultural change towards sharing between divisions Outturn costs available to encourage sharing

From Carrillo *et al.* (2006); reprinted by permission of the publisher, Taylor & Francis Ltd (http://www.informaworld.com).

The process is repeated for other issues to create or build a knowledge map which serves as a continuously evolving project memory for the organisation, identifying different knowledge sources, capturing and integrating new knowledge into the project knowledge base and knowledge tools to facilitate knowledge transfer to address specific problems. The elements on the knowledge map can be text, drawings, graphics, documents, directories, icons, symbols or models with links to more detailed knowledge. For example, a particular element could relate to the Skills Yellow Pages to provide a directory of experts. The knowledge map is very important but needs to be kept up-to-date to maintain its usefulness. Table 9.3 illustrates a range of knowledge issues associated with the different phases of PFI/PPP projects which an organisation may wish to explore. Similarly, detailed knowledge profiles could be developed for each specific stage of PFI/PPP projects.

Table 9.3 Examples of knowledge issues relating to knowledge map.

Phase	Examples of key knowledge issues
Planning and design development	Developing business case, selecting advisers for public sector, risk allocation, valuation of risks, determining value for money, evaluation of bids, design innovation, affordability
Construction	Project management structure, phasing and decanting
Service delivery and operation	Payment mechanism, monitoring service delivery, measuring and auditing performance, deductions for service failures

The knowledge map enables project team members in an organisation to learn from past and current projects through the navigation of information as well as the creation of new knowledge, by adding, refining and broadening the scope.

9.6   Creating an Action Plan for Learning and Capacity Building (Stage 3)

The final stage produces an action plan to implement a knowledge transfer and continuous improvement strategy. The project's industry collaborators were clear in stating that they required an action plan which provided a list of tasks to be undertaken as well as deadlines in which to complete the tasks. Thus, the action plan was devised with three main steps. These are as follows:

Step 3a: Identify tools and technologies required to support knowledge transfer

Step 3b: Identify appropriate monitoring mechanisms for knowledge transfer

Step 3c: Assess the organisation's readiness for knowledge transfer

Each of the above steps is supported by supplementary documents or detailed guidance notes in the form of appendices. Step 3a contains a matrix of knowledge transfer tools based on the Nonaka and Takeuchi's (1995) SECI (Socialisation, Externalisation, Combination and Internalisation) model discussed in Chapter 6. Research carried out on KM tools shows that organisations often have difficulties in identifying the most appropriate tools due to the large number of products in the marketplace and the overlap between in their function. The SeLEKT (*Searching* and *Locating Effective Knowledge Tools*) tool (Al-Ghassani *et al.*, 2005) is an integrated approach developed in a previous knowledge management research at Loughborough University to select the most appropriate tools based on the dimensions in the knowledge characteristics template shown in Figure 9.3. The knowledge transfer tools are also categorised according to 'entry level' tools and 'advanced level' tools to allow organisations flexibility in choosing appropriate tools to reflect their particular needs and level of maturity of their knowledge management strategy. This step also provides a glossary of terms to provide a better understanding of the tools available. Table 9.4 shows the matrix of the tools provided.

Step 3b provides a list of measures to monitor knowledge transfer and assess the effectiveness of learning and capacity building. As companies are at different levels of maturity in terms of knowledge transfer, there will be a need for different types of measures for monitoring knowledge transfer activities. This step therefore provides examples of 'entry level' and 'advanced level' measures depending on the type of knowledge transfer tool or technology used. The metrics are also categorised into individual, team and corporate metrics to allow appropriate selection for different constituents. An example of the measures used for knowledge transfer techniques is shown in Table 9.5.

Step 3c allows organisations to assess their readiness for knowledge transfer. It can be used either as a paper-based version or a Web-based version. The readiness assessment entails organisations completing a list of questions

Table 9.4   Stage 3 matrix of knowledge transfer tools.

Socialisation – *tacit to tacit*	Externalisation – *tacit to explicit*
Entry level:	Entry level:
Brainstorming	Best practice documents
Conferences/seminars/exhibitions	Databases
Face-to-face meetings	Discussion forum
Headhunting	Document archives
Mentoring	Skills Yellow Pages
Project reviews	
Succession planning	
Training	
Advanced level:	Advanced level:
Communities of practice	Expert systems
Video conferencing	Intelligent systems
Internalisation – *explicit to tacit*	Combination – *explicit to explicit*
Entry level:	Entry level:
Conferences/seminars/exhibitions	Intranets/extranets
Corporate universities	Best practice documents
Intranet/extranet	Procedure manuals
Search engines	
Succession planning	
Training	
Advanced level:	Advanced level:
Electronic document management systems	Data mining tools
Groupware	Document management systems
Virtual reality tools	

From Carrillo *et al.* (2006); reprinted by permission of the publisher, Taylor & Francis Ltd (http://www.informaworld.com).

Table 9.5   Stage 3 example measures for monitoring knowledge transfer.

Scope	Techniques	Examples of entry level measures	Examples of advanced level measures
Individual metrics	Mentoring	Frequency of meetings	Feedback (qualitative and quantitative)
	Conferences	Number of conferences	Evidence of positive impact/learning
Team metrics	Brainstorming	Frequency of sessions	Documentation and dissemination of session result
	Communities of practice	Number of active communities	Satisfaction survey of community members
Corporate metrics	Project reviews	Evidence that it occurs	Frequency Participation level Process change requests Lessons learned updates
	Succession planning	Evidence of succession planning	Evidence of structured action plan

From Carrillo *et al.* (2006); reprinted by permission of the publisher, Taylor & Francis Ltd (http://www.informaworld.com).

Organisational characteristics...

1. The organisation recognises the importance of sharing PFI knowledge

2. People are willing and motivated to share knowledge

3. Organisational/cultural issues such as power relations, barriers promoting 'silo' behaviour or knowledge hoarding have been addressed

4. There is a knowledge transfer strategy

5. There is a reward and incentive system for knowledge transfer

6. A change management programme has been implemented to facilitate knowledge transfer

7. Issues relating to confidentiality, copyright and reliability have been addressed

Figure 9.4 Organisational characteristics to assess readiness for knowledge transfer.

categorised into (1) organisational characteristics, (2) resource requirements and (3) results monitoring mechanisms. Examples of questions relating to organisational characteristics and the scoring mechanism are shown in Figure 9.4.

The inclusion of an organisational readiness assessment was considered important in helping to flag up issues that could have a detrimental impact on the company's knowledge transfer initiatives. Users are presented with a number of statements for which they have to respond using a Likert scale between 1 (strongly disagree) and 5 (strongly agree). The scoring system is based on the average score for each of the three categories. Scores less than 3.0 were considered poor (not ready), scores between 3.0 and 4.0 were considered fair (neutral) and scores over 4.0 were considered good (ready). The questions were evaluated by the project's industry collaborators to ensure their relevance and coverage. The outcome was a prioritisation of issues a company needs to address in order to improve knowledge transfer on PFI/PPP projects. The outputs highlight key areas of weaknesses that need to be addressed to ensure that knowledge transfer activities are not hindered by organisational factors. The organisational readiness assessment therefore allows companies to focus on particular problems and to improve specific aspects that need attention in order to successfully implement or improve knowledge transfer activities in an organisation. Low scores in the results sheet reflect the urgency associated with particular issues and the need to take action immediately.

Following the readiness assessment, the completion of the three steps results in an action plan for companies to implement. The action plan allows companies to:

- identify tasks to be undertaken to facilitate knowledge transfer to generate business opportunities;
- determine what tools and technologies are to be used to support these tasks;
- identify which knowledge transfer metrics should be used to assess the effectiveness of learning and capacity building;

- address issues highlighted in the organisational readiness assessment in terms of organisational reform and resources needed as well as results monitoring systems; and
- allocate named individuals for implementing knowledge transfer activities, with responsibility for monitoring progress within fixed timescales.

Table 9.6 shows a sample action plan form to implement learning and capacity building.

9.7   Evaluation of Framework

The framework was developed in phases and modified based on feedback at three workshops held with the project's industry collaborators to ensure that it met the project's objectives as well as the needs of industry. The first workshop was aimed at critiquing the conceptual model proposed and identifying issues and mechanisms for knowledge transfer. The first workshop involved five industry collaborators using a number of forms and guidance notes as supporting appendices to identify key issues. These included the following:

- A PFI process diagram to identify critical PFI processes requiring knowledge transfer
- A PFI transfer prioritisation form to narrow down the most relevant issues to address
- A PFI knowledge transfer template that explored the types of knowledge required, who were involved, current practices, scope for improvement and barriers to knowledge transfer

The feedback from the first workshop was used to fine-tune the development of the draft knowledge transfer framework. At the second workshop, the same five industry collaborators used the draft framework to address real issues that have arisen on PFI projects. Templates were provided for each of the three framework stages together with a list of tasks and guidance notes on completing each stage. The industry collaborators selected two examples that had commonality across the clients, engineering consultants and contractors. These were (1) benchmarking of PFI project data and (2) risk management. These examples were used in order to evaluate the framework's robustness, flow, consistency, gaps, and so on. The feedback from this workshop included the following:

- Need to provide more graphics to aid understanding of the flow across the various stages.
- Aspects of the framework were considered too lengthy and needed to be shortened and simplified.
- Illustrated examples should be provided to alert users to the type of input required.
- Example tools and technologies should be provided under the SECI matrix.

Table 9.6 Action plan form to implement learning and capacity building.

Stage 3 Tasks / 3.1 Restate type of knowledge	Improvement activities/tasks	Measures for monitoring knowledge transfer		Responsible person and position	Review/completion time/key date
Knowledge transfer solution		Entry level metrics	Advanced level metrics / Benchmarking		
3.2 Existing techniques to be improved (non-IT tools)	Review project performance on a regular basis	Number of quarterly project review meetings held	N/A at this point	A.N. Other, Managing Director	Quarterly reviews starting 20/04/09
3.3 New techniques required (non-IT tools)	Set up benchmarking group	Level of activity undertaken	N/A at this point	A.N. Other, Managing Director	Benchmarking group set up by 20/07/09
3.4 Existing technologies to be improved (IT tools)	Publish report status information	Number of hits on project database on intranet	N/A at this point	B.S. Brown, IT Manager	Quarterly reviews starting 20/04/09
3.5 New technologies required (IT tools)	Design, develop and implement an FM database	Number of hits on FM database	N/A at this point	B.S. Brown, IT Manager	FM database of costs completed by 20/01/10
Organisational readiness					
3.6 Identify organisational weakness to address in the short term	Establish process for identifying lessons learned on PFI projects		Lessons learned for three projects recorded and published on the intranet	C.A. Smith, Business Improvement Manager	Process in place by 20/04/09
3.7 Identify organisational weakness to address in the long term	Provide IT to support benchmarking		Access to benchmarking data within 60 seconds	B.S. Brown, IT Manager	Intranet pages available 20/01/10

From Carrillo et al. (2006); reprinted by permission of the publisher, Taylor & Francis Ltd (http://www.informaworld.com).

- Although the questions on organisational readiness assessment were found to be comprehensive and well structured, they needed (a) to identify which items were within the users' control and (b) a mechanism for highlighting the key issues more clearly.

These deficiencies were all addressed in the following ways:

- The original framework consisted of numerous forms with separate lists of tasks to complete supported by guidance notes and appendices. A flow chart was devised to graphically represent the different stages and their associated appendices. The guidance notes were condensed and placed on the page facing the form to be completed.
- Some of the forms were amalgamated, duplications removed and overall simplification of the flow between stages.
- Appendices provided worked examples of each stage using the workshop documentation to provide an aid for new users.
- The tools and technologies recommended were categorised into the SECI model and also according to entry or advanced level tools.
- The organisation readiness assessment was automated so that users would find it easier to select items under their control and also the results report used a traffic light system to highlight issues that were critical to address.

The third workshop was held to check that the changes proposed had been taken into consideration and to approve the final version of the framework and ensure that it was ready for dissemination. As a final check, one very experienced PFI industry collaborator was asked to examine the framework to ensure that both the framework and the guidance notes were sufficiently clear and relevant to industry's needs. This resulted in minor cosmetic changes to the Stage 3 form.

Feedback from the industry partners can be divided into two categories based on their level of PFI experience. Those collaborators with little PFI experience regarded the framework as providing ammunition for their line managers to adopt a more proactive approach to knowledge transfer and capacity building based on the results of the questionnaire survey, the case study reports and the knowledge transfer framework. The more experienced PFI collaborators saw it as a comprehensive and structured framework to encourage them to participate in knowledge transfer initiatives to improve their PFI portfolio.

9.8   Industrial Application, Dissemination and Benefits

The UK government's preferred way of procuring construction projects is the PFI/PPP procurement route. PFI/PPP projects are thus a valuable part of construction organisations' (contractors and consultants) project portfolio and there is a need to transfer knowledge between these projects to facilitate continuous improvement. The knowledge transfer framework developed is a learning and capacity building toolkit to encourage organisations to transfer knowledge between PFI/PPP projects. It was developed using different research methodologies. These included a (1) questionnaire survey of clients and construction companies (consultants and contractors) to investigate the current practices, participation, barriers and opportunities for PFI; (2) case studies of industry collaborators to understand PFI knowledge issues from the perspective of different stakeholders involved such as clients, SPV/SPC, designers and advisers, design and build contractors, facilities management contractors, their mechanisms for transferring knowledge on PFI projects and (3) action research with participating industrial collaborators to develop the conceptual model, apply the framework to addressing practical problems and to evaluate the usefulness of framework at key stages of its evolution and development.

There is a need for more structured approach to knowledge transfer practices to facilitate continuous improvement in industry. About three-quarters (76%) of client and construction organisation (consultants and contractors) agree that there is considerable scope for learning on PFI projects (Robinson *et al.*, 2004) but there is a lack of a structured framework in place to address knowledge transfer. The framework is therefore a powerful tool that is much needed by industry to improve their knowledge transfer practices. It allows organisations to identify key issues requiring knowledge transfer and to create an action plan, identifying key elements, resources and personnel, to transfer the knowledge required. The knowledge framework can also help to highlight other issues such as barriers to knowledge transfer, to judge the success of other knowledge transfer efforts using both low entry and advanced level measures.

9.9   Concluding Remarks

This chapter explored the opportunities on issues arising from PFI project and proposed a knowledge transfer framework that would enable organisations to be more proactive in managing and transferring knowledge on PFI projects. The framework consists of three stages that include (1) exploring PFI participation and opportunities, (2) mapping the organisation's knowledge and (3) creating an action plan for transferring knowledge. The knowledge transfer framework was evaluated using three workshops involving industry collaborators. The framework was found to be an appropriate way forward since it provides a structured way for identifying key issues, understanding what tools and technologies are available, and implementing and monitoring knowledge transfer tools and technologies on PFI projects. There is considerable scope for improvements in the PFI project delivery process. The knowledge transfer framework presented in this chapter will enable both construction organisations and clients to improve their current practices and reap the attendant benefits.

References

Adams, J. (2005) United we stand. *PFI Intelligence Bulletin* 10(8), 12–13.
Al-Ghassani, A.M. (2003) *Improving the Structural Design Process: A Knowledge Management Approach*. PhD Thesis. Loughborough University.

Al-Ghassani, A.M., Anumba, C.J., Carrillo, P.M., and Robinson, H.S. (2005) Tools and technologies for knowledge management. In: Anumba, C.J., Egbu, C., and Carrillo, P.M. (eds). *Knowledge Management in Construction*. Blackwell, Oxford, pp. 83–102.

Argote, L., and Ingram, P. (2000) Knowledge transfer: a basis for competitive advantage in firms. *Organizational Behaviour and Human Decision Processes* 82(1), 150–169.

Argote, L., Ingram, P., Levine, J., and Moreland, R. (2000) Knowledge transfer in organizations: learning from the experiences of others. *Organizational Behaviour and Human Decision Processes* 82(1), 1–8.

Audit Commission (2003) *PFI in Schools: The Quality and Cost of Buildings and Services Provided by Early Private Finance Initiative Schemes*. Audit Commission, London.

Bresnen, M., and Marshall, N. (2000) Partnering in construction: a critical review of issues, problems and dilemmas. *Construction Management and Economics* 18(2), 227–237.

Brooking, A. (1996) *Intellectual Capital – Core Asset for the Third Millennium Enterprise*. International Thompson Business Press, London.

Carrillo, P., Robinson, H., Foale, P., Anumba, C., and Bouchlaghem, D. (2008) Participation, barriers and opportunities in PFI: the United Kingdom experience. *Journal of Management in Engineering* 24(3), 138–145.

Carrillo, P.M., Robinson, H.S., Al-Ghassani, A.M., and Anumba, C.J. (2004) Knowledge management in UK construction: strategies, resources and barriers. *Project Management Journal* 35(1), 46–56.

Carrillo, P.M., Robinson, H.S., Anumba, C.J., and Bouchlaghem, N.M. (2006) Knowledge transfer framework: the PFI context, *Construction Management and Economics* 2410, 1045–1056.

Crawley, L.G., and Karim, A. (1995) Conceptual model of partnering. *ASCE Journal of Management in Engineering* 11(5), 33–39.

Ernst and Young (2002) *Progress and Prospects: A Survey of Healthcare PFI*. Available online at http://www.ey.com (Accessed 5 February 2003).

Gann, D. (2001) Putting academic ideas into practice: technological progress and the absorptive capacity of construction organisations. *Construction Management and Economics* 19(3), 321–330.

Goh, S.C. (2002) Managing effective knowledge transfer: an integrative framework and some practical implications. *Journal of Knowledge Management* 6(1), 23–30.

HM Treasury (2004) *Value for Money Assessment Guidance*. HM Treasury, London.

Levy, M., Loebbecke, C., and Powell, P. (2001) *SMEs, Co-Opetition and Knowledge Sharing: The IS Role*. Global Co-operation in the New Millennium. Proceedings of the 9th European Conference on Information Systems. Bled, Slovenia.

McAdam, R., and McCreedy, S. (1999) The process of knowledge management within organizations: a critical assessment of both theory and practice. *Knowledge and Process Management* 6(2), 101–113.

NAO (2007) *Improving the PFI Tendering Process*. HC149. The Stationery Office, London.

Newell, S., Scarbrough, H., Robertson, M., and Swan, J. (2002) *Managing Knowledge Work*. Palgrave Global Publishing, Basingstoke, Hampshire.

Nonaka, I., and Takeuchi, H. (1995) *The Knowledge-Creating Company: How Japanese Companies Create the Dynamics of Innovation*. Oxford University Press, New York.

Robinson, H.S., Carrillo, P.M., Anumba, C.J., and Al-Ghassani, A.M. (2001) *Linking Knowledge Management Strategy to Business Performance in Construction*

Organisations. Proceedings of the 17th Annual ARCOM Conference, 2. Salford University, 5 September 2001, pp. 577–586.

Robinson, H.S., Carrillo, P.M., Anumba, C.J., and Bouchlaghem, N.M. (2004) *Investigating Current Practices, Participation and Opportunities in the Private Finance Initiative*. Internal Report. Loughborough University.

Szulanski, G. (2000) The process of knowledge transfer: a diachronic analysis of stickiness. *Organizational Behaviour and Human Decision Processes* 82(1), 9–27.

von Krogh, G., Nonaka, I., Aben, M. (2001) Making the most of your company's knowledge: a strategic framework. *Long Range Planning* 34(4), 421–439.

Conclusion

Various studies in developed and developing countries have shown that there is a significant shortfall in infrastructure investment and lack of maintenance resulting in a deteriorating stock of public infrastructure capital to support the delivery of core public services. Public-private partnership is an approach that is increasingly adopted to facilitate the improvement of public services where there are public sector budgetary constraints and there is a need for innovation by stimulating private investment in infrastructure facilities such as health, education, transport, custodial, defence and social housing, regeneration and waste management. The traditional procurement approach requiring upfront investment from the public sector is often seen as fragmented because design, construction, operational activities or functions are separated. This has sometimes resulted in poorly performing or dysfunctional buildings delivering poor services. The alternative public-private partnership is a whole life or integrated approach from design to facilities management and service delivery aimed at addressing the problems associated with the traditional approach by creating a shift in emphasis from 'building contracting and lump sum payment' to 'service contracting and performance-based payment'. However, it is important that appropriate policy, strategic and implementation structures and processes are in place to address the key objectives of the public sector in PFI/PPP projects. The World Bank (2007, p. 3) defines a successful public-private partnership (PPP) programme as one that provides the following:

- The services a government needs
- Offers value for money as measured or compared with public service provision including the cost of bearing risk
- Complies with general standards of good governance and specific government policy

Another critical success factor that should be added is sustainability to reflect the increasing need to balance economic objectives with environmental and social obligations. This book has addressed critical policy, strategic and implementation issues in Chapters 2 and 3 to ensure that appropriate PPP projects are brought forward that reflect actual services required by the government or public sector as client. At the heart of the UK PFI model is the output specification defining the need and services required by the public sector/government, which is discussed in detail in Chapter 3. Preparing the output specification requires a systematic planning approach underpinned by robust processes and controls to support and define the services a government or public sector client needs, to develop the business case for a project and to assess its viability. The output specification has two elements to it, both of which are critical in determining the capital and operating costs (whole life cost) of a PPP/PFI project and the payment received by the operator. The output specification (accommodation *standard*) relates to the physical condition of the buildings and the output specification (*performance standard*) relates to range of soft and hard facilities management services required and how well the operator carries out the services it is responsible for. A well-drafted output specification is therefore fundamental to the successful delivery of long-term services required by the government or public sector in PFI/PPP projects. The book has demonstrated the importance of understanding the key elements of the output specification in the planning and development of PFI/PPP projects. The output-based approach in PFI/PPP projects goes beyond the provision of capital assets by focusing on 'what' services are required by the public sector client in contrast to the traditional approach which focuses on 'how' a building should be delivered (technical specification). The output specification also provides the basis for the comparing traditional procurement cost (the public sector comparator (PSC) benchmark cost) with private sector PFI/PPP bid.

In return for the investment in design, construction of facilities and the services provided by the private sector, public sector client pays an annual payment (unitary charge) to the SPV/SPC based on performance monitoring to ensure that value for money is achieved. The payment reflects operational or production risks transferred to the private sector relating to the accommodation standard (availability of facilities) and facilities management services required in the output specification. The link between output specification defining the needs of the public sector and the payment mechanism therefore provides an incentive for the private sector to deliver services that the public sector needs and to ensure that value for money is maintained throughout the concession agreement.

The implementation processes, principles and application of governance discussed in Chapters 3, 4 and 5 provide powerful controls to ensure the output specification is appropriately developed and value for money is achieved through effective relationship between public and private sector, adequate risk transfer, proper valuation of risk and comparison with the traditional procurement approach, the project is affordable and the selection of an appropriate bidder. PFI/PPP projects have raised awareness of risk in ways that traditional public procurement has not been able to do. In order to achieve the 'value for money objectives' in service delivery, the public and private sector partners need to reach a mutually acceptable risk allocation strategy before the contract is awarded. Although it is impossible to eliminate all project risks in planning and design development, construction, and operation and service delivery, the governance tools described in Chapter 4 provide an effective mechanism for risk allocation and management. Performance monitoring and measurement mechanisms linked to the output specification ensure that operational risks are tested and the financial penalties (deductions) for availability and service failures are applied to maintain value for money. Other processes and tools such as Gateway Review process, project management, early warning systems and decision-making structures act as effective controls to ensure that project will deliver services that the government or public sector needs in compliance with the policy and strategic framework and implementation processes discussed in Chapters 2 and 3. This enables due processes to be followed, actors and decision-makers to comply with the general principles of good project governance to provide value for money for the public sector client. Further chapters included on knowledge management principles, knowledge transfer case studies, capacity building and knowledge transfer framework reinforce the need to be able to capture lessons learnt from earlier PPP/PFI projects to 'feed forward' to future projects. Understanding governance and knowledge management principles are therefore crucial to facilitate continuous improvement in PPP projects.

10.1   Governance Issues

The term 'governance' in public sector organisations commonly relates processes to deliver to services (e.g. clinical services as in health care organisations), and to manage financial and staff resources and is used to mean 'assurance'. This book focuses on the management processes that support PFI/PPP projects and are critical to speeding up the process of decision-making and moving a PFI/PPP project forward to completion. Good governance through a controlled planning and procurement process is integral to achieving the project outcomes based on the needs of the public sector, and excellence in design quality, construction and service delivery based on the output specification. However, in the absence of controls, projects are in danger of running over time and incurring increase in expenditure on planning, design, construction costs including the cost of external advisers. The absence of good governance to speed up the long gestation period between design and construction leaves many PFI/PPP projects vulnerable to downstream design changes as public sector organisations undergo modernisation in their practices due to regulations, codes of practice and as technology changes. A key lesson learnt is that organisations must ensure that their governance framework is in alignment with the project size and complexity. A further lesson learnt is that whilst governance tools to support PFI/PPP projects are readily available, these are of little value without knowledge management the provision of appropriate training as to 'when' and 'how' to use the tools effectively. It is therefore necessary to consider the implications for knowledge management, training and capacity building in developing a governance framework to improve project delivery.

Organisations involved in PFI/PPP schemes should develop a governance structure that is aligned to the appropriate organisational objectives with clearly defined roles in terms of decision-making, responsibilities and accountability. Organisations must understand the importance of the key elements of governance (people, roles, processes, standards, tools and mechanisms for project compliance) to ensure continuous improvement in project delivery. In particular, there is a need within the governance framework for clarity between a person's 'role' and their associated 'responsibilities'. Project managers must demonstrate that they understand and are trained to use governance tools such as project control and risk management systems for effective monitoring of projects. The independent Gateway Review process is an example of a major governance tool. However, Gateway Review teams must have the authority to bring a PFI/PPP project to a temporary halt until the SRO (Senior Responsible Officer) delivers the recommendations within the timeframe allocated for change. Lessons learnt should be made accessible to project directors of future PFI projects.

10.2   Knowledge Management and Capacity Building Issues

The broad objective of knowledge management is to accelerate national and local capacity building particularly in the public and private sectors to improve their knowledge in the planning and design development, construction and operational phases of PFI/PPP projects. Transferring lessons learnt is crucial in PFI/PPP projects to improve policy, strategic framework, implementation processes and to achieve good governance. As a result of knowledge management, various changes, controls, best practices and standards have been introduced in the PFI/PPP process such as streamlining the procurement processes, the development of standard guidance documents, standard contract documents, application of design toolkit and BREEAM, batched schemes or bundling of projects to reduce high bidding/transaction costs, lengthy negotiation periods which are key barriers. However, there is still a considerable scope for knowledge sharing, learning and capacity building.

Capacity building strategies permeate the different phases and stages of the PFI/PPP projects and all sectors where PFI/PPP is currently applied. As a result of knowledge management and capacity building strategies, the implementation of PFI/PPP in the United Kingdom has improved significantly and lessons learned from previous projects are now being applied in other sectors such as social housing, urban regeneration, waste, and so on. In the United Kingdom and other developed countries, problems of capacity building have often been addressed through flexible labour and immigration policies, allowing highly skilled migrants to participate in the labour market to alleviate problems in design and construction industry. However, the situation is more difficult in developing and transition economies with limited skill base as they often rely heavily on technical assistance funded by bilateral and multilateral development agencies or paid external advisers to embark on PFI/PPP programmes. There is recognition that given the UK's considerable experience, they can export PPP/PFI expertise in other European countries and developing/middle-income countries with aspirations or enabling legislation to improve public services through a PPP framework.

The training of PPP specialists and the time required to accumulate experience may take years. There were initial problems of capacity in the United

Kingdom in the early stages of PFI/PPP as only the larger contractors were able to participate in PPP/PFI projects due to complex procurement and significant transaction costs associated with bidding. There is evidence also to suggest that problems of capacity have threatened the PFI/PPP market and the achievement of value for money particularly in the health sector with complex hospital projects. The implementation of PPP programmes in developing countries, without a careful evaluation of the resource implications, can create major problems as they often have to rely on international PPP contractors and consultants. Careful consideration is therefore required in formulating policies and strategies for implementing PPP/PFI projects in developed as well as developing countries. Knowledge management and knowledge transfer programmes are crucial in building capacity; otherwise PPP projects will be derailed and those that are implemented will remain very expensive. Significant timescales may therefore be required to develop or address capacity problems.

The knowledge management principles and knowledge transfer issues discussed in Chapters 6, 7 and 8 as well as the practical knowledge management tools described in Chapter 9 in the book are based on the findings of recent research projects. The tools are particularly useful for industry practitioners such as PPP/PFI consultants, contractors and client organisations as they will facilitate the implementation of knowledge management strategy, the evaluation and monitoring of learning and knowledge transfer activities to build capacity for PFI/PPP projects.

10.3   Sustainability of PFI/PPP Projects

Both governance and knowledge management play an essential role in the sustainability of PFI/PPP projects in terms of the economic, social and environmental dimensions. The availability of good governance tools supporting key processes such as project initiation, the development of business cases, risk management, decision-making, accountability and approval processes will ensure that project costs and expenditure are effectively controlled. Governance will also enable the selection and implementation of appropriate projects designed to meet genuine public sector needs, pass the 'value for money' and 'risk transfer' tests and more significantly to deliver services that address key social objectives in health, education, prisons, transport, social housing and other sectors. The impact of knowledge management is substantial in the UK PFI/PPP programme and will no doubt continue to be a significant factor in the reduction of costs associated with PFI/PPP projects. This includes cost such as transaction costs, bidding and negotiation costs, training costs, costs associated with wrong decision-making, delay in making decisions, planning and design costs due to streamlining of the procurement process, better controls, standardisation of PPP/PFI documents, elimination of mistakes, better understanding of risks and the introduction of innovation, whether incremental or major.

The environmental dimension of sustainability remains a major issue as buildings and structures contribute significantly to global warming. There is an intrinsic link between the whole life cost which is at the heart of the PFI/PPP models and sustainability. PFI through the use of whole life appraisal techniques is changing the approach to design, and the early involvement of the FM specialist means that design solutions are increasingly linked to social and environmental performance targets. PFI/PPP projects have led to the increased awareness of whole life issues and the role of FM in early design, the significance of asset management to improve energy performance and other social and environmental performance indicators in buildings. There are strong incentives for PFI/PPP contractors to respond to the challenge of improving the sustainability of buildings through designing, constructing and maintenance as regular payments and unitary charges are linked to service performance. Incorporating sustainability objectives in PFI/PPP projects can help in improving material wastage, water and energy usage, air quality, safety and security.

Reference

Sustainable Development Department in East Asia and Pacific, The World Bank (2007) *Public Private Partnership Units: Lessons for Their Design and Use in Infrastructure*. The World Bank and Public-Private Infrastructure Advisory Facility (PPIAF).

Appendix A: An Example of Output Specification (Accommodation Standard)

A.1.1   Purpose and scope

The standard wards will provide the majority of in-patient accommodation for both Trusts. This document describes a 32-bed ward, which includes 7 larger bed spaces for Level 2 patients. All beds will be allocated on a speciality basis and will be arranged in a combination of 4-bed bays and single rooms.

To provide a flexible facility to support:

- All inpatient services requiring a general level of clinical support
- Patients needing single organ system monitoring and support
- Level 2 patients requiring more detailed observation or intervention in-cluding those stepping up, or down, from higher levels of care, that is Level 3 (intensive care)

Exclusions:

- Adults who require Level 3 care
- Children

A.1.2   Service trends

Patients being referred in the future are likely to require more complex types of treatments, as the more routine work will be increasingly undertaken by local district general hospitals. This is likely to lead to a higher proportion of more dependent patients with a longer length of stay than that is currently seen, and therefore clinical areas should be designed to allow for flexible management of patients.

A.1.3   Workload activity and facility numbers

13 × 32 bed standard bed wards.
Each ward to have the following:

- 8 single rooms – of which 1 is an isolation room and 3 are Level 2
- 6 × 4 bed bays – of which 1 is Level 2

A.1.4   People

Maximum volumes

Functional area	Patients	Staff	Visitors	Total
Reception area	4	3	2	9
Waiting/sitting room	8	0	8	16
Bed space	1	3	2	5
Bed space Level 2	1	4	2	7
Staff base (per 4 beds)	0	4	0	4
Treatment room	1	4		5

Total staff numbers

	Total staff	Numbers requiring access to staff rest	Numbers requiring a change locker	Numbers requiring access to lockers
Nurses	50	4	50	0
Clinicians	7	2	0	0
Admin	3	1	0	3
Total	**60**	**7**	**50**	**3**

A.1.5   Work patterns

24 hours per day, 7 days a week.

A.1.6   Access

Security

All ward entrances will have proximity card security entry system for staff and the main patient/visitor entrance will have a videophone entry system for all other visitors.

Staff access

The entrance to allow staff access to their rest and change area must not be the main ward entrance.

Within the ward, the following rooms will be staff-only access by proximity card security entry:

- Clean utility
- Dirty utility
- Drug preparation
- Treatment rooms
- Equipment store
- Offices and meeting rooms
- Staff rest and change

Patient access

Each ward must have a nominated single main entrance for patient- and visitor-access only.

Patients/visitor access will only gain access when a member of staff operates the video entry door control system.

Patients will arrive by foot, or on trolley, chair or bed.

A.1.7   Patient and staff flows

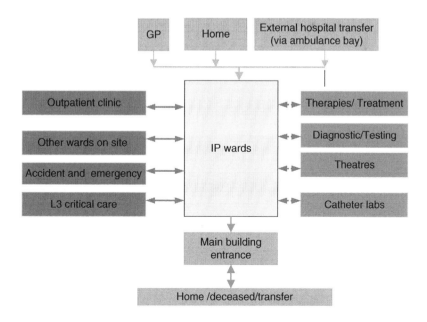

A.1.8   External key adjacencies

Adjacency to	Location (see key)	Reason	Essential/ desirable
To other adult wards so that all standard wards are located in the same area of each hospital build	3	So that all wards are located in a 'block' facilitates better patient management	Essential
Intensive care unit (Level 3)	3	To facilitate ease of transfer of acute patients requiring Level 3 care	Essential
Theatres/recovery	4	To facilitate safe transfer of post-operative patients	Essential
Accident and emergency	4	To facilitate the safe transfer of emergency patients	Essential
Imaging	4	High volume of ward patients will need to access this service	Essential
Inpatient therapy areas (one department in both Trust buildings)	4	Ward patients will need to access this service for rehabilitation	Essential

Location key: 1 – Directly adjacent; 2 – Same floor; 3 – Adjacent floor (horizontal or vertical); and 4 – Same building.

A.1.9   Key design principles

The key design principles must be read in conjunction with the following documents:

- M&E Matrix
- Infection Control Document
- Clinical Planning Exemplar Text

Internal key adjacencies

Generic:

1. The ward reception will be in a prominent position to ensure all visitors are directly visible on entry to the ward. The ward reception must not be isolated from the main ward area.
2. The visitors waiting area should be collocated with the ward reception; it must be away from the main clinical area to protect patient privacy and reduce noise levels.

3. The interview room must be located to ensure that distressed relatives can make a dignified exit from the ward.
4. Staff support areas, treatment room, drugs room, clean and dirty utilities must be collocated with good access to all bed areas.
5. The patient sitting room must be in a central location for reasons of visibility, access and patient safety.
6. Sisters' office to be positioned away from the main clinical area.
7. Visitor's toilet to be collocated with the visitors wait.
8. Staff rest/change is shared between two directly adjacent wards; the design must allow for the rest area to be central, allowing staff from both wards to be physically close to their clinical area.
9. 'Back of house' activities should be carried out in areas not observed by patients:
   - Movement of bodies from the inpatient area to the mortuary
   - Distribution, handling, storage and collection of food, laundry, stores and supplies
   - Movement of staff prior to and following their shift
   - Staff rest, change, WC

Overarching department design

1. All beds will be wired to facilitate Level 2 care.
2. HDU beds must be grouped together in one area.
3. Location of HDU beds must allow the shortest distance between these beds and the local core patient bed lifts.
4. Staff bases and room arrangements should facilitate maximum observation of inpatient beds and resuscitation equipment, pneumatic tube point, and drug prep room.
5. Direct visibility of the head of the bed into single rooms is essential; screening must also allow this to be obscured to facilitate patient privacy. This allows staff to observe the patient without disturbing him or her unnecessarily. It also allows the patient to view outside of the room, and so reduce the feeling of isolation.
6. Dual controlled blinds inpatient bed areas and treatment rooms must be included to provide privacy.
7. Each 4 – bedroom must have a dedicated en suite shower/WC with access immediately outside the bay entrance.
8. Location of disposal hold must allow this store to be emptied without the need to enter the main ward while still allowing staff to fill the hold from within.
9. The resuscitation equipment must be central to the ward area.

A.1.10   Functional content

Department/room	Number of rooms	Comment
Ward area per 32 beds		
4-bed bay and en suite assisted shower/WC	5	
Single room and en suite assisted shower/WC	4	
4-bed bay (Level 2) and en suite disabled shower/WC	1	
Single room (Level 2 and en suite disabled shower/WC	3	
Isolation single room and en suite assisted shower/WC	1	
Lobby to isolation room	1	
Staff work base	8	One per 4-bed bay, one for four single rooms
Support accommodation per 32 beds		
Reception		
Clean utility		
Drug prep room		
Sitting room/waiting room		
Interview room		
Assisted bathroom		
Assisted WC		
WC		
Treatment room		
Dirty utility		
Pneumatic tube station		
Resus trolley bay		
Pantry		
Bev Bay		
Food trolley hold		
Equipment store		
Sisters office – 2 person		
MDT work station		
Hoist bay		
Linen bay		
Staff WC		
Cleaners room		
Shared accommodation between 2 × 32 bed wards		
Local vending		
Disabled WC/baby change		
Therapies room		
Mobile x-ray bay		
Specialist nurse office – 2 person		
Specialist nurse office – 4 person		
Modern matron office		
MDT resource/discussion room		
Staff rest inc. bev bay		
Disposal hold		
Cook chill regeneration kitchen		
Staff change (male and female)		

Appendix B1: Output Specification (Facilities Management Standard)

NHS Standard Service Level Specifications

Service-Specific Specification – Car Parking

This document is the result of a joint collaboration between NHS Estates and the Private Finance and Investment Branch of the Department of Health.

B1.1   Definitions

Any reference to 'this service level specification' in this Part [ ] shall be a reference to this car parking service level specification (including the appendices hereto).

In this service level specification, the following words and phrases shall have the following meaning:

'Car parking areas'	means all car parks and all other areas designated for parking including on-road parking for all types of vehicles including but not limited to cars, bicycles, and so on.
'Car parking users'	means patients, staff and bona fide visitors on Trust-related business.
'Priority staff'	means those staff authorised by the Trust allocations committee staff parking rights.

B1.2   Key Objectives

Project Co shall provide a comprehensive car parking service including traffic management across the Trust site(s). The service shall be operable 24 hours per day 365(6) days per year on a planned and ad hoc basis. Project Co shall:

(a)   provide a secure and safe car park environment for patients, staff and bona fide visitors to the hospital, their vehicles and their property;

(b)   provide car parking areas that maximise the use of the space whilst minimising the risk of crime and pollution;

(c)   provide Trust traffic management across the Trust site(s) to ensure the free flow of traffic ensuring access to the facilities at all times;

(d) provide an administration service that controls all parking-related administration and revenue collection; and

(e) promote the NHS Green Transport Plan by encouraging the use of sustainable
transport modes.

B1.3   Key Customers

The key customers for this service are as follows:

(a) Patients
(b) Priority staff
(c) Staff
(d) Visitors
(e) Emergency services
(f) Traffic/transport department
(g) Service providers/contractors

B1.4   Process

B1.4.1   Scope   PP Ref.

B1.4.1.1   Project Co shall comply with all requirements set out in subpart C
(General Service Specification) of Part [1] of this Schedule [14]
relevant to the delivery of the car parking service.

B1.4.1.2   In addition to the applicable provisions set in the General Service
Specification, Project Co shall comply with the service standards
and service requirements of this service-specific specification.

B1.4.1.3   Project Co shall provide the following services and elements, as
part of the car parking service so as to meet the service standards:

(a) Traffic management
(b) Car parking areas including the following:
   (i) Equipment
   (ii) Designated/priority parking
   (iii) Maintenance issues
(c) Car park administration including the following:
   (i) Revenue collection and accounting
   (ii) Complaint processing
   (iii) Permit system

B1.4.1.4   In addition to the general policies detailed in the General Service   SP01
Specification, Project Co shall comply with the following
service-specific policies and regulations including but not limited
to the following:

(a) Aggressive and violent behaviour policy
(b) Car parking policy and procedures
(c) Security policy

(d)   Trust security manual
(e)   NHS NAHAT security manual
(f)   Road traffic regulations
(g)   Fire policy
(h)   NHS Green Transport Plan
(i)   Aggressive and violent behaviour policy
(j)   Association of Chief Police Officers Secured Car Parking

B1.4.2   Minimum service requirements   PP Ref.

01   Project Co shall provide the car parking service 24 hours a day   **SP02**
     365(6) days per year on a planned and reactive basis as defined in
     the response and rectification times described in section 'Table
     B1'.

Traffic management

02   Project Co shall keep all entrances, exits and internal roadways   **SP03**
     within the Trust site(s) clear from vehicles and other obstructions,
     thus maintaining free flow of traffic at all times. These
     responsibilities include but are not limited to the following:

     (a)   Enforced removal of such obstructions at Project Co's
           expense
     (b)   Dealing with customer complaints in accordance with Trust
           complaints policy
     (c)   Administration of parking fines and penalties, if appropriate
     (d)   Acting at all times in a courteous and polite manner
     (e)   The implementation and management of a traffic control
           system for use during a major accident, bomb alert or other
           major incident at the hospital

03   Project Co shall provide traffic marshalling duties, if required by   **SP04**
     the Trust, during special occasions/visits organised by the Trust.

04   Project Co shall ensure owners of vehicles displaying incorrect or   **SP05**
     out-of-date permits are identifiable and traceable and
     appropriate action must be taken in accordance with Trust policy
     and the service level specification.

Car park areas

05   Project Co shall provide, maintain, operate and replace when   **SP06**
     necessary, access and egress equipment, mechanical or
     otherwise, to ensure car park areas are used by patients, staff,
     permit holders and/or bona fide visitors only. Such control
     measures shall minimise the potential for causing congestion in
     so doing and shall have sufficient capacity to cope with peak
     traffic flow.

06   Project Co shall provide, maintain, operate and replace when   **SP07**
     necessary, revenue collection equipment, mechanical or
     otherwise, to ensure charges are collected on behalf of the Trust.
     This shall include but not be limited to the following:

(a) Collection of revenue
(b) Replenishment of consumables such as tickets, receipts, and so on
(c) Displaying current parking charges and car park regulations

07	Project Co shall ensure all equipment and machinery are commissioned, operated and maintained in good safe working order and in accordance with manufacturer's instructions and requirements of the appropriate portable appliance regulations at all times.	SP08
08	Project Co shall ensure payment, and access/egress control mechanisms are of a suitable design for use by all users including disabled drivers.	SP09

Designated/priority parking

09	Project Co shall provide designated and priority parking areas within the car park areas and ensure that all designated spaces are used by their intended user group only. The quantity of each type is to be agreed with the Trust representative. Project Co shall provide designated spaces for the following user groups:	SP10

(a) Patients
(b) Priority staff
(c) Staff
(d) Disabled
(e) Young mothers
(f) Elderly
(g) Emergency services
(h) Drop-off and delivery areas
(i) Bicycles and motorcycles

For avoidance of doubt, this shall include all pubic and private type vehicles for the specified user groups.

10	Project Co shall ensure that all space dimensions are in accordance with BS5810 and HBN40. Parking spaces for those with disabilities or young mother parking shall take full regard of extra space requirements of wheelchairs, pushchairs and/or vehicles with loading ramps and be situated near to public entrances/exits.	SP11

Car park maintenance

11	Project Co shall regularly inspect the fabric and fittings of the car park areas and internal site roadways and report any damage to the help desk promptly. Such damage may include but not be limited to the following:	SP12

(a) Damaged car park or road surface
(b) Curbing and footpaths
(c) Overhanging or obtrusive vegetation
(d) Inadequate street and/or car park lighting
(e) Road or space definition markings
(f) Inadequate or damaged signage

Car park management and administration

12	The charging levels will be agreed with the Trusts on an annual basis. The payment system will be based upon a daily charge for visitors, and the issue and payment for an annual permit for staff. Charges will need to take account of the Green Transport Policy to be agreed with the Trust. Project Co shall not change the rates payable by users of the car parks without first gaining written consent from the designated Trust representative.	SP13
13	Project Co shall ensure all parking rates and regulations, including the policy for lost tickets and money, shall be displayed at every entrance to the car park and at each payment station.	SP14
14	Project Co shall ensure all staff, patients and visitors comply with the payment system agreed with the Trust.	SP15
15	Project Co shall implement, maintain and administer a comprehensive parking permit system to:	SP16

(a) provide the necessary equipment and systems to maintain a permit administration database to maintain a record of all permits issued and transactions;

(b) issue and reclaim permits/keycards and other concessionary devices in accordance with the Trust policy; and

(c) ensure that permits are dispatched within 24 hours of receipt of the Trust-approved applications.

16	Project Co shall develop and maintain a system for recording and acting on customer feedback and satisfaction in line with Trust policy. Car parking user customer satisfaction questionnaires are to be carried out (twice yearly) in a format agreed with the Trust representative.	SP17
17	Project Co shall record and respond to enquire within a suitable timescale, including the resolution of day-to-day car parking problems.	SP18
18	Project Co shall provide to the Trust the following reports:	SP19

(a) Ticket sales
(b) Permits issued and reclaimed
(c) All infringements
(d) All incidents of crime within the car park(s)

19	Project Co shall develop and manage a car share scheme on behalf of the Trust. The scheme shall be free to all staff.	SP20
20	Project Co shall provide advice, help and guidance to vehicle owners, which shall include parking arrangement directing car park users to available parking spaces advising of costs, and so on.	SP21
21	Project Co shall proactively assist car park users in the event of punctures, flat batteries, lights left on, and so on.	SP22
22	Project Co may withdraw parking concessions from any person violating the parking regulations and Trust parking policy. This shall only be implemented with written agreement from the designated Trust representative.	SP23

For the avoidance of doubt, money accrued from staff parking permits shall be passed onto the Trust. Project Co shall advice if a guaranteed income will be provided from the patient/visitor car parking.

| 23 | Project Co shall ensure all car park management staff are fully trained in first aid, control and restraint, management of violence and aggression and cardiopulmonary resuscitation. | SP24 |

Security

| 24 | Project Co shall maintain the car parks at such a level as to achieve and retain for the duration of the contract the Assoc. of Chief Police Officer's Secured Car Park Award. | SP25 |
| 25 | The car park management staff will effectively support the security service in responding to incidents and requests for assistance. | SP26 |

B1.5   Table B1

B1.5.1   Response and rectification

For the purpose of determining response times and rectification times, the failure or request for service shall be categorised as emergency, urgent or routine. Table B.1 provides the relevant definitions:

Table B.1   Failure or request for service categories.

Category	Definition
Emergency	Any events felt to be life threatening or serious enough to cause significant harm or damage. Any request for a service which is required to avoid a life-threatening event or an event serious enough to cause significant damage or disruption
Urgent	Any faults that shall cause operational problems if not attended to quickly, or which may develop into an emergency if not remedied. Any request for a service which requires attendance quickly to avoid operational problems, or an emergency if not remedied
Routine	Any faults that are not seen as immediately detrimental and not causing significant operational problems. Any request for a service that is not seen as immediately detrimental and not causing significant operational problems if not attended to

Table B.2   Response and rectification time.

Category	Maximum response time	Maximum rectification time
Emergency	Immediate (5 minutes)	15 minutes
Urgent	30 minutes	15 minutes
Routine	1 hour	30 minutes

Response and rectification times run concurrently.

Appendix B2: Example of Performance Parameters (Car Park)

B2.1 Performance Parameters

Reference	Performance parameter	SF type	Category	Response	Rectification	Measurement period	Monitoring method
SP01	Service specific Trusts and statutory requirements and standards; service level specification and method statement are complied with at all times	QF	High	N/A	N/A	M	1, 2, 4, 7, 8
SP02	The car parking service is manned 24 hours a day 365(6) days per year	QF	Medium	N/A	N/A	M	2, 4, 5, 8
SP02	Emergency requests are responded to in accordance with response times and service level specification	FE	A–E	Immediate (5 minutes)	N/A	PR	1, 4, 8
SP02	Urgent requests are responded to in accordance with response times and service level specification	FE	A–C	30 minutes	N/A	PR	1, 4, 8
SP02	Routine requests are responded to in accordance with response times and service level specification	FE	A–C	1 hour	N/A	PR	1, 4, 8
SP02	Emergency requests are accurately rectified in accordance with the rectification times and service level specification	FE	A–E	N/A	15 minutes	PR	1, 4, 8
SP02	Urgent requests are accurately rectified in accordance with the rectification times and service level specification	FE	A–C	N/A	15 minutes	PR	1, 4, 8
SP02	Routine requests are accurately rectified in accordance with the rectification times and service level specification	FE	A–C	N/A	30 minutes	PR	1, 4, 8

Reference	Performance parameter	SF type	Category	Response	Rectification	Measurement period	Monitoring method
Traffic management							
SP03	Controls are in place to ensure that internal roadways are kept clear at all times	FE	A–E	1 hour	30 minutes	PR	1, 4, 8
SP03	All 'no parking' or restricted parking areas are to be kept free of unauthorised vehicles or other obstructions	FE	A–E	1 hour	30 minutes	PR	1, 4, 8
SP04	Traffic marshalling duties are provided as required by the Trust	QF	Low	N/A	N/A	PR	1, 4, 8
SP05	Adequate permit tracing and tracking facilities are in operation, with appropriate actions taken in the event of vehicles displaying incorrect/out-of-date permits	QF	Low	N/A	N/A	PR	1, 4, 8
Car park access							
SP06	Systems and/or equipment is in place to ensure satisfactory admission to and exit from the car parks without causing congestion	QF	High	N/A	N/A	D	1, 2, 4, 8
SP07	Revenue collection systems and equipment are in place, operable and stocked with appropriate level of consumables where applicable	QF	High	N/A	N/A	D	1, 2, 4, 8
SP08	All equipment related to the provision of the car parking service have been correctly commissioned and are fully operable in accordance with manufacturer's instructions	QF	Medium	N/A	N/A	M	1, 4, 8
SP09	Revenue collection and access/egress equipment are suitable for use by all car park users including disabled car parking users	QF	High	N/A	N/A	M	1, 4, 7, 8

Reference	Performance parameter	SF type	Category	Response	Rectification	Measurement period	Monitoring method
Designated/priority parking							
SP10	Dedicated spaces and drop-off points clearly marked and systems are in place, operable and are used by their intended user only	QF	Medium	N/A	N/A	D	1, 4, 8
SP11	Parking spaces comply with BS5810 and HBN40	QF	High	N/A	N/A	B	1, 4, 8
Car park maintenance							
SP12	A system of regular inspections is operable and all faults are recorded with the help desk promptly in the agreed manner	QF	Medium	N/A	N/A	M	1, 2, 3, 4, 5, 8
Car park management and administration							
SP13	Parking charges are/have been agreed with the Trust representative and are correctly charged at the agreed level	QF	High	N/A	N/A	M	1, 2, 3, 4, 5, 8
SP14	Parking charges and parking regulations are clearly displayed in the agreed format at every car park entrance and revenue collection point	QF	Low	N/A	N/A	W	1, 2, 3, 4, 8
SP15	All car park users hold a valid ticket or permit	QF	High	N/A	N/A	D	1, 4, 8
SP16	A central record of all permits issued past and present is in operation and the data are correct	QF	Medium	N/A	N/A	M	2, 3, 4, 8
SP16	All permits/key cards, and so on, are issued and reclaimed in accordance with Trust policy	QF	Medium	N/A	N/A	M	2, 3, 4, 8
SP16	Permits are dispatched within 24 hours of receipt of the Trust-approved application	QF	Medium	N/A	N/A	PR	1, 2, 3, 4, 8

Reference	Performance parameter	SF type	Category	Response	Rectification	Measurement period	Monitoring method
SP17	A system for recording customer feedback and satisfaction is operational	QF	Medium	N/A	N/A	M	2, 3, 4, 6, 8
SP17	Customer satisfaction is above 90%	QF	Medium	N/A	N/A	B	6
SP18	Enquires are recorded and adequately responded to within a suitable timescale	QF	Low	N/A	N/A	PR	1, 4, 5, 6, 8
SP19	The monthly report is submitted to the designated Trust representative on the agreed date in the agreed format and quality	QF	Low	N/A	N/A	M	2, 4, 8
SP20	A car share scheme is operational and available to all staff	QF	Low	N/A	N/A	B	2, 4, 8
SP21	Accurate advice is provided to car park users as and when requested	QF	Low	N/A	N/A	M	1, 4, 6, 8
SP22	Assistance is provided to vehicle owners in the event of puncture, flat batteries, lights left on, and so on	QF	Medium	N/A	N/A	M	1, 4, 6, 8
SP23	Parking concessions are withdrawn from car park users once agreement has been granted by the Trust	QF	Low	N/A	N/A	M	1, 3, 4, 8
SP24	All car park management staff are suitably trained as per service level specification	QF	Medium	N/A	N/A	M	2, 4, 5
Security							
SP25	The car park has current Association of Chief Police Officer's Secured Car Parking Award	QF	Medium	N/A	N/A	A	7
SP26	The car park management staff will, as required, effectively support the security service	QF	Medium	N/A	N/A	M	1, 2, 4, 8

Please note that the performance parameters will require reviewing once the payment mechanism has been reviewed and drafted.

B2.2   Key Performance Indicators

KPI reference	Key performance indicator	Performance range		
		Green	Amber	Red
K01	Number of complaints per month	< [] number	[] – [] number	> [] number
K02	Number of crime incidents per month	< [] number	[] – [] number	> [] number

Appendix C: Performance Measurement System

Performance measurement works on several levels. Each service is subdivided into scopes which are then further divided into aspects of service.

Service

Scopes

Aspects

Each aspect has one or more performance standards and performance standards are measured in one of three ways: measurement, tariff or audit

---

The performance score and aspect score are subject to a weighting criterion.

Actual performance score × weighting = standard weighted score (total 100)
Standard weighted score × aspect weighting = aspect weighted score (total 40)
Aspect weighted score (added for all applicable headings) = scope score
Scope score divided by aspect weighting (added for all applicable headings) × 100 gives scope score as a %

Each scope is then given an importance rating based on the following:

Level 4 – Utmost (this scope has a major impact on the running of the Trust)

Level 3 – Primary importance (this scope has a significant impact on the service unit/department's operation)

Level 2 – Very important (this scope affects the service units/department's operation)

Level 1 – Important (this scope has a lesser effect on the service unit/department)

The % scope score is then adjusted by the importance rating to give the adjusted scope score (based on the following criteria).

Level 4 (scope score must achieve 97.5% and above for the following):

Examples
Facilities
Catering
Security
Materials management

Level 4 (scope score must achieve 95% and above for the following):

Examples
Domestics
Portering and waste management
Car parking and traffic control
Laundry
Telecommunications

If the scores are equal to or greater than 97.5% and 95%, respectively, the % scope score will be adjusted to 100%.

Failure to achieve the respective scope scores will be subject to the following adjustment:

Scope score less than 97.5% – scope score% will be reduced to 87.5% (except facilities).
Scope score less than 95% – scope score% will be reduced to 85%.

Facilities management service is broken into two groups:
Group A – Mechanical, electrical, specialist, PMG, infrastructure and building maintenance

Failure to achieve 97.5% on these categories means the scope score% will be reduced to 75%.

Group B – Space planning, small works and projects, registration/documentation and landscaping

Failure to achieve 97.5% on these categories means the scope score% will be reduced to 75%.

The service performance score for facilities service is then calculated as the average of the scores for the two groups.

Level 3
Where a scope score has an importance rating of Level 3, the performance score will be the scope score.

Level 2
Where a scope score has an importance rating of Level 2, the performance score will be calculated as follows:

Scope score% + [(100% − Scope score%) divided by 2]

Level 1
Where a scope score has an importance rating of Level 1, the performance score will be calculated as follows:

Scope score% + [(100% − Scope score%) × 2/3]

The overall service performance score for the month is calculated by multiplying all the scope scores (with importance ratings) together for that measurement month.

Performance band

Service perform score	Service perform score	Perform band	Deficiency points
(Facilities, catering, security, materials management)	(Domestic, Laundry Telecoms, Pottering Waste Management, Car Parking & Traffic Control)		
100–97.5%	100–95%	1	0
Less than 97.5–94.5%	Less than 95–92.5%	2a	1
Less than 94.5–92.5%	Less than 92.5–90%	2b	2
Less than 92.5–90%	Less than 90–87.5%	3a	6
Less than 90–87.5%	Less than 87.5–85%	3b	8
Less than 87.5–84%	Less than 85–82.5%	4a	11
Less than 84%	Less than 82.5%	4b	16

It should be noted that rectification periods are allowed contractually, and provided these are carried out accordingly, the performance standard scores will not be affected.

Rectification periods allowed (based on importance level)

Level		Frequency of task			
Annually	Twice or more	Once daily	Weekly	Monthly	Annually
	Daily (minutes)	(minutes)	(hours)	(days)	(weeks)
4	10 minutes	20 minutes	4 hours	1 day	1 week
3	20 minutes	40 minutes	6 hours	2 days	2 weeks
2	30 minutes	60 minutes	12 hours	3 days	3 weeks
1	60 minutes	120 minutes	24 hours	4 days	4 weeks

Index

Printed and bound by CPI Group (UK) Ltd, Croydon, CR0 4YY